by Steven Holzner, PhD

Wiley Publishing, Inc.

Physics II For Dummies®

Published by
Wiley Publishing, Inc.
111 River St.
Hoboken, NJ 07030-5774
www.wiley.com

WILEY

About the Author

Steven Holzner taught Physics at Cornell University for more than a decade, teaching thousands of students. He's the award-winning author of many books, including *Physics For Dummies, Quantum Physics For Dummies,* and *Differential Equations For Dummies,* plus *For Dummies* workbooks for all three titles. He did his undergraduate work at MIT and got his PhD from Cornell, and he has been on the faculty of both MIT and Cornell.

Dedication

To Nancy, of course.

Author's Acknowledgments

The book you hold in your hands is the product of many people's work. I'd particularly like to thank Acquisitions Editor Tracy Boggier, Senior Project Editor Alissa Schwipps, Senior Copy Editor Danielle Voirol, Technical Editors Laurie Fuhr and Ron Reifenberger, and the many talented folks in Composition Services.

Publisher's Acknowledgments

We're proud of this book; please send us your comments at http://dummies.custhelp.com. For other comments, please contact our Customer Care Department within the U.S. at 877-762-2974, outside the U.S. at 317-572-3993, or fax 317-572-4002.

Some of the people who helped bring this book to market include the following:

Acquisitions, Editorial, and Media Development

Senior Project Editor: Alissa Schwipps

Acquisitions Editor: Tracy Boggier

Senior Copy Editor: Danielle Voirol

Contributor: Neil Clark

Assistant Editor: Erin Calligan Mooney

Senior Editorial Assistant: David Lutton

Technical Editors: Laurie Fuhr, Ron Reifenberger

Senior Editorial Manager: Jennifer Ehrlich

Editorial Assistants: Rachelle Amick, Jennette ElNaggar

Cover Photos: © Thorsten | Dreamstime.com

Cartoons: Rich Tennant (www.the5thwave.com)

Composition Services

Project Coordinator: Katherine Crocker

Layout and Graphics: Carrie A. Cesavice, Amy Hassos, Mark Pinto, Erin Zeltner

Proofreader: Tricia Liebig

Indexer: Sharon Shock

Publishing and Editorial for Consumer Dummies

Diane Graves Steele, Vice President and Publisher, Consumer Dummies

Kristin Ferguson-Wagstaffe, Product Development Director, Consumer Dummies

Ensley Eikenburg, Associate Publisher, Travel

Kelly Regan, Editorial Director, Travel

Publishing for Technology Dummies

Andy Cummings, Vice President and Publisher, Dummies Technology/General User

Composition Services

Debbie Stailey, Director of Composition Services

Contents at a Glance

Table of Contents

Introduction

For many people, physics holds a lot of terror. And Physics II courses do introduce a lot of mind-blowing concepts, such as the ideas that mass and energy are aspects of the same thing, that light is just a mix of electric and magnetic fields, and that every electron zipping around an atom creates a miniature magnet. In Physics II, charges jump, light bends, and time stretches — and not just because your instructor lost the class halfway through the lecture. Throw some math into the mix, and physics seems to get the upper hand all too often. And that's a shame, because physics isn't your enemy — it's your ally.

The ideas may have come from Albert Einstein and other people who managed to get laws and constants and units of measurement named after them, but you don't have to be a genius to understand Physics II. After all, it's only partially rocket science — and those are ultra-cool, nearing-the-speed-of-light rockets.

Many breakthroughs in the field came from students, researchers, and others who were simply curious about their world, who did experiments that often didn't turn out as expected. In this book, I introduce you to some of their discoveries, break down the math that describes their results, and give you some insight into how things work — as physicists understand it.

About This Book

Physics II For Dummies is for the inquiring mind. It's meant to explain hundreds of phenomena that you can observe all around you. For example, how does polarized light really work? Was Einstein really right about time dilation at high speeds? Why do the electromagnets in electric motors generate magnetism? And if someone hands you a gram of radioactive material with a half-life of 22,000 years, should you panic?

To study physics is to study the world. *Your* world. That's the kind of perspective I take in this book. Here, I try to relate physics to your life, not the other way around. So in the upcoming chapters, you see how telescopes and microscopes work, and you find out what makes a properly cut diamond so brilliant. You discover how radio antennas pick up signals and how magnets make motors run. You see just how fast light and sound can travel, and you get an idea of what it really means for something to go radioactive.

When you understand the concepts, you see that the math in physics isn't just a parade of dreadful word problems; it's a way to tie real-world measurements to all that theory. Rest assured that I've kept the math in this book relatively simple — the equations don't require any knowledge beyond algebra and trigonometry.

Physics II For Dummies picks up where a Physics I course leaves off — after covering laws of motion, forces, energy, and thermodynamics. Physics I and Physics II classes have some overlap, so you do find info on electricity and magnetism in both this book and in *Physics For Dummies*. But in *Physics II For Dummies*, I cover these topics in more depth.

A great thing about this book is that *you* decide where to start and what to read. It's a reference you can jump into and out of at will. Just head to the table of contents or the index to find the information you want.

Conventions Used in This Book

Some books have a dozen stupefying conventions that you need to know before you can start reading. Not this book. All you need to know is the following:

- New terms are given in italics, like *this,* and are followed by a definition.

- Variables, like *m* for *mass,* are in italics. If you see a letter or abbreviation in a calculation and it isn't italicized, you're looking at a unit of measurement; for instance, 2.0 m is 2.0 meters.

- Vectors — those items that have both a magnitude and a direction — are given in bold, like this: *B.*

And those are all the conventions you need to know!

What You're Not to Read

Besides the main text of the book, I've included some extra little elements that you may find enlightening or interesting: sidebars and paragraphs marked with Technical Stuff icons. The sidebars appear in shaded gray boxes, and they give you some nice little examples or tell stories that add a little color or show you how the main story of physics branches out. The Technical Stuff paragraphs give you a little more technical information on the matter at hand. You don't need this to solve problems; you may just be curious.

If you're in a rush, you can skip these elements without hurting my feelings. Without them, you still get the main story.

Foolish Assumptions

In this book, I assume the following:

✔ You're a student who's already familiar with a Physics I text like *Physics For Dummies*. You don't have to be an expert. As long as you have a reasonable knowledge of that material, you'll be fine here. You should understand ideas such as mass, velocity, force, and so on, even if you don't remember all the formulas.

✔ You're familiar with the metric system, or SI (the International System of Units). You can convert between units of measurement, and you understand how to use metric prefixes. I include a review of working with measurements in Chapter 2.

✔ You know basic algebra and trigonometry. I tell you what you need in Chapter 2, so no need to worry. This book doesn't require any calculus, and you can do all the calculations on a standard scientific calculator.

How This Book Is Organized

Like physics itself, this book is organized into different parts. Here are the parts and what they're all about.

Part 1: Understanding Physics Fundamentals

Part I starts with an overview of Physics II, introducing the goals of physics and the main topics covered in a standard Physics II course. This part also brings you up to speed on the basics of Physics I — just what you need for this book. You can't build without a foundation, and you get the foundation you need here.

Part II: Doing Some Field Work: Electricity and Magnetism

Electricity and magnetism are a big part of Physics II. Over the years, physicists have done a great job of explaining these topics. In this part, you see both electricity and magnetism, including info on individual charges, AC (alternating current) circuits, permanent magnets, and magnetic fields — and perhaps most importantly, you see how electricity and magnetism connect to create electromagnetic waves (as in light).

Part III: Catching On to Waves: The Sound and Light Kinds

This part covers waves in general, as well as light and sound waves. Of the two, light is the biggest topic — you see how light waves interact and interfere with each other, as well as how they manage when going through single and double slits, bouncing off objects, passing through glass and water, and doing all kinds of other things. The study of optics includes real-world objects such as lenses, mirrors, cameras, polarized sunglasses, and more.

Part IV: Modern Physics

This part brings you into the modern day with the theory of special relativity, the particle-wave duality of matter, and radioactivity. Relativity is a famous one, of course, and you see a lot of Einstein in this part. You also see many other physicists who chipped in on the discussion of matter's travels as waves. You read all about radioactivity and atomic structure, too.

Part V: The Part of Tens

The chapters in this part cover ten topics in rapid succession. You take a look at ten physics experiments that changed the world, leading to discoveries in everything from special relativity to radioactivity. You also look at ten online calculators that can assist you in solving physics problems.

Icons Used in This Book

You find icons in this book, and here's what they mean:

This icon marks something to remember, such as a law of physics or a particularly important equation.

Tips offer ways to think of physics concepts that can help you better understand a topic. They may also give you tips and tricks for solving problems.

This icon means that what follows is technical, insider stuff. You don't have to read it if you don't want to, but if you want to become a physics pro (and who doesn't?), take a look.

Where to Go from Here

In this book, you can jump in anywhere you want. You can start with electricity or light waves or even relativity. But if you want the full story, start with Chapter 1. It's just around the corner from here. Happy reading!

If you don't feel comfortable with the level of physics taken for granted from Physics I, check out a Physics I text. I can recommend *Physics For Dummies* wholeheartedly.

Part I

Understanding Physics Fundamentals

The 5th Wave   By Rich Tennant

"This is my old physics teacher, Mr. Wendt, his wife Doris, and their two children, Quark and Wormhole.".

In this part . . .

In this part, you make sure you're up to speed on the skills you need for Physics II. You start with an overview of the topics I cover in this book. You also review Physics I briefly, making sure you have a good foundation in the math, measurements, and main ideas of basic physics.

Chapter 1

Understanding Your World: Physics II, the Sequel

In This Chapter

▶ Looking at electricity and magnetism

▶ Studying sound and light waves

▶ Exploring relativity, radioactivity, and other modern physics

*P*hysics is not really some esoteric study presided over by guardians who make you take exams for no apparent reason other than cruelty, although it may seem like it at times. Physics is the human study of *your* world. So don't think of physics as something just in books and the heads of professors, locking everybody else out.

Physics is just the result of a questioning mind facing nature. And that's something everyone can share. These questions — what is light? Why do magnets attract iron? Is the speed of light the fastest anything can go? — concern everybody equally. So don't let physics scare you. Step up and claim your ownership of the topic. If you don't understand something, demand that it be explained to you better — don't assume the fault is with you. This is the human study of the natural world, and you own a piece of that.

Physics II takes up where Physics I leaves off. This book is meant to cover — and unravel — the topics normally covered in a second-semester intro physics class. You get the goods on topics such as electricity and magnetism, light waves, relativity (the special kind), radioactivity, matter waves, and more. This chapter gives you a sneak preview.

Getting Acquainted with Electricity and Magnetism

Electricity and magnetism are intertwined. Electric charges in motion (not static, nonmoving charges) give rise to magnetism. Even in bar magnets, the tiny charges inside the atoms of the metal cause the magnetism. That's why you always see these two topics connected in Physics II discussions. In this section, I introduce electricity, magnetism, and AC circuits.

Looking at static charges and electric field

Electricity is a very big part of your world — and not just in lightning and light bulbs. The configuration of the electric charges in every atom is the foundation of chemistry. As I note in Chapter 14, the arrangement of electrons gives rise to the chemical properties of matter, giving you everything from metals that shine to plastics that bend. That electron setup even gives you the very color that materials reflect when you shine light on them.

Electricity studies usually start with electric charges, particularly the force between two charges. The fact that charges can attract or repel each other is central to the workings of electricity and to the structure of the atoms that make up the matter around you. In Chapter 3, you see how to predict the exact force involved and how that force varies with the distance separating the two charges.

Electric charges also fill the space around them with electric field — a fact familiar to you if you've ever felt the hairs on your arm stir when you've unloaded clothes from a dryer. Physicists measure electric field as the force per unit charge, and I show you how to calculate the electric field from arrangements of charges.

Next up is the idea of *electric potential,* which you know as *voltage.* Voltage is the work done per unit charge, taking that charge between two points. And yes, this is exactly the kind of voltage you see stamped on batteries.

With those three quantities — force, electric field, and voltage, you nail down static electric charges.

Moving on to magnetism

What happens when electric charges start to move? You get magnetism, that's what. *Magnetism* is an effect of electric charge that's related to but distinct from the electric field; it exists only when charges are in motion. Give an electron a push, send it sailing, and presto! You've got magnetic field. The idea that moving electric charges cause magnetic field was big news in physics — that fact's not obvious when you simply work with magnets.

Electric charges in motion form a *current*, and various arrangements of electric current create different magnetic fields. That is, the magnetic field you see from a single current-bearing wire is different from what you see from a loop of current — let alone a whole bunch of loops of current, an arrangement known as a *solenoid*. I show you how to predict magnetic field in Chapter 4.

Not only do moving electric charges give rise to magnetic fields, but magnetic fields also affect moving electric charges. When an electric charge moves through a magnetic field, that charge feels a force on it at right angles to the magnetic field and the direction of motion. The upshot is that left to themselves, moving charges in uniform magnetic fields travel in circles (an idea chemists appreciate, because that's what allows a mass spectrometer to sort out the chemical makeup of a sample). How big is the circle? How does the radius of the circle correlate with the speed of the charge? Or with the magnitude of the charge? Or with the strength of the magnetic field? Stay tuned. The answers to all these questions are coming up in Chapter 4.

AC circuits: Regenerating current with electric and magnetic fields

Students often meet electrical circuits in Physics I (you can read about simple direct current [DC] circuits in *Physics For Dummies*). In Chapter 5, you get the Physics II version: You take a look at what happens when the voltage and current in a circuit fluctuate in time in a periodic way, giving you *alternating voltage* and *currents*. You also encounter some new circuit elements, the inductor and capacitor, and see how they behave in AC circuits. Many of the electrical devices that people use every day depend on such elements in alternating currents.

In reading about the inductor, you also encounter one of the fundamental laws that relates electric and magnetic fields: Faraday's law, which explains how a changing magnetic field induces a voltage that generates its own magnetic field. This law doesn't just apply to inductors; it applies to all electric and magnetic fields, wherever they occur in the universe!

Riding the Waves

Waves are a huge topic in Physics II. A *wave* is a traveling disturbance that carries energy. If the disturbance is *periodic,* the amount of disturbance repeats in space and time over a distance called the *wavelength* and a time called the *period.* Chapter 6 delves into the workings of waves so you can see the relationships among the wave's speed, wavelength, and *frequency* (the rate at which cycles pass a particular point). In the rest of the chapters in Part III of this book, you explore particular types of waves, including electromagnetic waves (such as light and radio waves) and sound.

Getting along with sound waves

Sound is just a wave in air, and the various interactions of sound waves are just a result of the behaviors shared by all waves. For instance, sound waves can reflect off a surface — just let sound waves collide with walls and listen for the echo. Sound waves also interfere with other waves, and you can hear the effects — or silence, as the case may be. These two kinds of interaction form the basis for understanding the harmonic tones in music.

The qualities of a sound, such as pitch and loudness, depend on the properties of the wave. As you may have noticed by hearing the change of pitch of a siren on a police car as it passes by, pitch changes when the source or the listener moves. This is called the *Doppler effect.* You can take this to the extreme by examining the shock wave that happens when objects move very quickly through the air, breaking the sound barrier. This is the origin of the sonic boom. I cover all this and more in Chapter 7.

Figuring out what light is

You focus on light a good deal in Physics II. How light works is now well-known, but that wasn't always the case. Imagine the excitement James Clerk Maxwell must've felt when the speed of light suddenly jumped out of his equations and he realized that by combining electricity and magnetism, he'd come up with light waves. Before that, light waves were a mystery — what made them up? How could they carry energy?

After Maxwell, all that changed, because physicists now knew that light was made up of electrical and magnetic oscillations. In Chapter 8, you follow in Maxwell's footsteps to come up with his amazing result. There, you see how to calculate the speed of light using two entirely different constants having to do with how well electric and magnetic fields can penetrate empty space.

As a wave, light carries energy as it travels, and physicists know how to calculate just how much energy it can carry. That amount of energy is tied to the magnitude of the wave's electric and magnetic components. You get a handle on how much power that light of a certain intensity can carry in Chapter 8.

Of course, light is only the visible portion of the *electromagnetic spectrum* — and it's a small part at that. All kinds of electromagnetic radiation exist, classified by the frequency of the waves: X-rays, infrared light, ultraviolet light, radio waves, microwaves, even ultra-powerful gamma waves.

Reflection and refraction: Bouncing and bending light

Light's interaction with matter makes it interesting. For instance, when light interacts with materials, some light is absorbed and some reflected. This process gives rise to everything you see around you in the daily world.

Reflected light obeys certain rules. Primarily, the *angle of incidence* of a light ray — that is, the angle at which the light strikes the surface (measured from a line pointing straight out of that surface) — must equal the *angle of reflection* — the angle at which the light leaves the surface. Knowing how light is going to bounce off objects is essential to all kinds of devices, from the periscopes in submarines to telescopes, fiber optics, and even the reflectors that the Apollo astronauts placed on the moon. Chapter 10 covers the rules of reflection.

Light can also travel through materials, of course (or people wouldn't have windows, sunglasses, stained glass, and a lot more). When light enters one material from another, it bends, a process known as *refraction* — which is a big topic in Chapter 9. The amount the light bends depends on the materials involved, as measured by their *indexes of refraction*. That's useful to know in all kinds of situations. For example, when lens-makers understand how light bends when it enters and leaves a piece of glass, they can shape the glass to produce images. You take a look through lenses next.

Searching for images: Lenses and mirrors

If you're eager to look at the practical applications of Physics II topics, you'll probably enjoy optics. Here, you work with lenses and mirrors, allowing you to explore the workings of telescopes, cameras, and more.

Focusing on lenses

Lenses can focus light, or they can diverge it. In either case, you can get an image (sometimes upright, sometimes upside down, sometimes bigger than the object, sometimes smaller). The image is either virtual or real. In a *real image,* the light rays converge, so you can put a screen at the image location and see the image on the screen (like at the movies). A *virtual image* is an image from which the light appears to diverge, such as an image in a magnifying glass.

Armed with a little physics, you have the lens situation completely under control. If you're visually inclined, you can find info on the image using your drawing skills. I explain how to draw ray diagrams, which show how light passes through a lens, in Chapter 9.

You can also get numeric on light passing through lenses. The thin-lens equation gives you all you need to know here about the object and image, and you can even derive the magnification of lenses from that equation. So given a certain lens and an object a certain distance away, you can predict exactly where the image will appear and how big it will be (and whether it'll be upside down or not).

If one lens is good, why not try two? Or more? After all, that's the idea behind microscopes and telescopes. You get the goods on such optical instruments in Chapter 9, and if you want, you can be designing microscopes and telescopes in no time.

All about mirrors/srorrim tuoba llA

You can get numeric on the way mirrors reflect light, whether a mirror is flat or curved. For instance, if you know just how much a mirror curves and where an object is with reference to the mirror, you can predict just where the image of the object will appear.

In fact, you can do more than that — you can calculate whether the image will be upright or upside down. You can calculate just how high it will be compared to the original object. You can even calculate whether the image will be real (in front of the mirror) or virtual (behind the mirror). I discuss mirrors in Chapter 10.

Calling interference: When light collides with light

Not only can light rays interact with matter; they can also interact with other light rays. That shouldn't sound too wild — after all, light is made up of electric and magnetic components, and those components are what interact with the electric fields in matter. So why shouldn't those components also interact with similar electric and magnetic components from other light rays?

When the electric component of a light ray is at its maximum and it encounters a light ray with its electric component at a minimum, the two components cancel out. Conversely, if the two light rays happen to hit just where the electric components are at a maximum, they add together. The result is that when light collides with light, you can get *diffraction* patterns — arrangements of light and dark bands, depending on whether the net result is at a maximum or minimum. In Chapter 11, you see how to calculate what the diffraction patterns look like for an assortment of different light sources, all of which has been borne out by experiment.

Branching Out with Modern Physics

The 20th century saw an explosion of physics topics, and collectively, those topics are called *modern physics*. Some revolutionary ideas — such as quantum mechanics and Einstein's theory of special relativity — changed the foundations of how physicists saw the universe; Isaac Newton's mechanics didn't always apply. As physicists delved deeper into the workings of the world, they found more and more powerful ideas, which allowed them to describe exponentially more about the world. This led to developments in technology, which meant that experiments could probe the universe ever more minutely (or expansively).

Most people have heard of relativity and radioactivity, but you may not be familiar with other topics, such as *matter waves* (the fact that when matter travels, it exhibits many wave-like properties, just like light) or *blackbody radiation* (the study of how warm objects emit light). I introduce you to some of these modern-physics ideas in this section.

Shedding light on blackbodies: Warm bodies make their own light

If you've ever seen an incandescent light bulb at work (or you've glanced at the sun), you know that hot things emit light. In fact, any body with any warmth at all emits electromagnetic waves, such as light.

In particular, physicists can calculate the wavelength of the electromagnetic waves where the emitted spectrum peaks, given an object's temperature. This topic is intimately tied up with *photons* — that is, particles of light — and you can predict how much energy a photon carries, given its wavelength. Details are in Chapter 13.

Speeding up with relativity: Yes, $E = mc^2$

Here it is at last: special relativity and Einstein. What, exactly, does $E = mc^2$ mean? It means that matter and energy can be considered interchangeable, and it gives the energy-equivalent of a mass m at rest. That is, if you have a tomato that suddenly blows up, converting all its mass into energy (not a likely event), you can calculate how much energy would be released. (***Note:*** Converting 100 percent of a tomato's mass into pure energy would create a huge explosion; a nuclear explosion converts only a small percentage of the matter involved into energy.)

Besides $E = mc^2$, Einstein also predicted that at high speeds, time stretches and length contracts. That is, if a rocket ship passes you traveling at 99 percent of the speed of light, it'll appear contracted along the direction of travel. And time on the rocket ship passes more slowly than you'd expect, using a clock at rest with respect to you. So if you watch a rocket ship pass by at high speed, do clocks tick more slowly on the rocket ship than they do for you, or is that some kind of trick? No trick — in fact, the people on the rocket ship age more slowly than you do, too.

Airplanes travel at much slower speeds, but the same effect applies to them — and you can calculate just how much younger a jet passenger is than you (but here's a disappointing tip to people searching for the fountain of youth: It's an immeasurably small amount of time). You explore special relativity in Chapter 12.

Assuming a dual identity: Matter travels in waves, too

Light travels in waves — that much doesn't take too many people by surprise. But the fact that matter travels in waves can be a shocker. For example, take your average electron, happily speeding on its way. In addition to exhibiting particle-like qualities, that electron also exhibits wave-like qualities — even so much so that it can interfere with other electrons in flight, just as two light rays can, and produce actual diffraction patterns.

And electrons aren't the only type of matter that has a wavelength. Everything does — pizza pies, baseballs, even tomatoes on the move. You wrap your mind around this when I discuss matter waves in Chapter 13.

Meltdown! Knowing the αβγ's of radioactivity

Nuclear physics has to do with, not surprisingly, the nucleus at the center of atoms. And when you have nuclear physics, you have radioactivity.

In Chapter 15, you find out what makes up the nucleus of an atom. You see what happens when nuclei divide *(nuclear fission)* or combine *(nuclear fusion)* — and in particular, you see what happens when nuclei decay by themselves, a process known as *radioactivity*.

Not all radioactive materials are equally radioactive, of course, and *half-life* — the time it takes for half of a sample to decay — is one good measure of radioactivity. The shorter the half-life, the more intensely radioactive the sample is.

You encounter all the different types of radioactivity — alpha, beta, and gamma — in the tour of the subject in Chapter 15.

Chapter 2

Gearing Up for Physics II

This chapter prepares you to jump into Physics II. If you're already a physics ace, there's no need to get bogged down here — just fly into the physics topics themselves, starting with the next chapter. But if you're not fast-tracked for the physics Nobel Prize, it wouldn't hurt to scan the topics here, at least briefly. Doing so can save you a lot of time and frustration in the chapters coming up.

Math and Measurements: Reviewing Those Basic Skills

Physics excels at measuring and predicting the real world, and those predictions often come though math. So to be a physicsmeister, you have to have certain skills down cold. And because this is Physics II, I assume that you're somewhat familiar with the world of physics and some of those basics already. You look at those skills here in refresher form (if you're unclear about anything, check out a book like *Physics For Dummies* (Wiley) to get up to speed).

The following skills are pretty basic; you can't get through Physics I without them. But make sure you have at least a passing acquaintance with the topics in this section — especially if it's been quite some time since Physics I.

Using the MKS and CGS systems of measurement

The most common measurement systems in physics are the centimeter-gram-second (CGS) and meter-kilogram-second (MKS) systems. The MKS system is more common. For reference, Table 2-1 lists the primary units of measurement, along with their abbreviations in parentheses, for both systems.

Table 2-1	Metric Units of Measurement	
Type of Measurement	*CGS Unit*	*MKS Unit*
Length	Centimeters (cm)	Meters (m)
Mass	Grams (g)	Kilograms (kg)
Time	Seconds (s)	Seconds (s)
Force	Dynes (dyn)	Newtons (N)
Energy (or work)	Ergs (erg)	Joules (J)
Power	Ergs/second (erg/s)	Watts (W) or joules/second (J/s)
Pressure	Baryes (Ba)	Pascals (Pa) or newtons/square meter (N/m^2)
Electric current	Biots (Bi)	Amperes (A)
Magnetic field	Gausses (G)	Teslas (T)
Electric charge	Franklins (Fr)	Coulombs (C)

These are the primary measuring sticks that physicists use to measure the world with, and that measuring process is where physics starts. Other measuring systems, such as the foot-pound-second (FPS) system, are around as well, but the CGS and MKS systems are the main ones you see in physics problems.

Making common conversions

Measurements don't always come in the units you need them in, so doing physics can involve a lot of conversions. For instance, if you're using the meter-kilogram-second system (see the preceding section), you can't plug measurements in centimeters or feet into your formula — you need to get them in the right units first. In this section, I show you some values that are equal to each other and an easy way to know whether to multiply or divide when doing conversions.

Looking at equal units

Converting between CGS (centimeter-gram-second) and MKS (meter-kilogram-second) units happens a lot in physics, so here's a list of equal values of MKS and CGS units for reference — come back to this as needed:

- **Length:** 1 meter = 100 centimeters
- **Mass:** 1 kilogram = 1,000 grams
- **Force:** 1 newton = 10^5 dynes
- **Energy (or work):** 1 joule = 10^7 ergs
- **Pressure:** 1 pascal = 10 barye
- **Electric current:** 1 ampere = 0.1 biot
- **Magnetism:** 1 tesla = 10^4 gausses
- **Electric charge:** 1 coulomb = 2.9979×10^9 franklins

Converting back and forth between MKS and CGS systems is easy, but what about other conversions? Here are a some handy conversions that you can come back to as needed. First, for length:

- 1 meter = 1,000 millimeters
- 1 inch = 2.54 centimeters
- 1 meter = 39.37 inches
- 1 mile = 5,280 feet = 1.609 kilometers
- 1 kilometer = 0.62 miles
- 1 angstrom (Å) = 10^{-10} meters

Here are some conversions for mass:

- 1 slug (foot-pound-second system) = 14.59 kilogram
- 1 atomic mass unit (amu) = 1.6605×10^{-27} kilograms

These are for force:

- 1 pound = 4.448 newtons
- 1 newton = 0.2248 pounds

Here are some conversions for energy:

- 1 joule = 0.7376 foot-pounds
- 1 British thermal unit (BTU) = 1,055 joules

> ✔ 1 kilowatt-hour (kWh) = 3.600×10^6 joules
>
> ✔ 1 electron-volt = 1.602×10^{-19} joules

And here are conversions for power:

> ✔ 1 horsepower = 550 foot-pounds/second
>
> ✔ 1 watt = 0.7376 foot-pounds/second

Using conversion factors: From one unit to another

If you know that two values are equal to each other (see the preceding section), you easily use them to convert from one unit of measurement to another. Here's how it works.

First note that when two values are equal, you can write them as a fraction that's equal to 1. For instance, suppose you know that there are 0.62 miles in a kilometer:

$$1 \text{ km} = 0.62 \text{ miles}$$

You can write this as

$$\frac{1 \text{ km}}{0.62 \text{ mi.}} = 1 \quad \text{or} \quad \frac{0.62 \text{ mi.}}{1 \text{ km}} = 1$$

Each of these fractions is a *conversion factor*. If you need to go from miles to kilometers or kilometers to miles, you can multiply by a conversion factor so that the appropriate units cancel out — without changing the value of the measurement, because you're multiplying by something equal to 1.

For instance, suppose you want to convert 30 miles to kilometers. First, write 30 miles as a fraction:

$$\frac{30 \text{ miles}}{1}$$

Now you need to multiply by a conversion factor. But which version of the fraction do you use? Here, *miles* is in the numerator, so to get the *miles* to cancel out, you want to multiply a fraction that has *miles* in the denominator. Because $\frac{1 \text{ km}}{0.62 \text{ mi.}} = 1$, you can multiply 30 miles by that fraction without changing the measurement. Then the *miles* on the bottom cancels the *miles* on the top:

$$\frac{30 \ \cancel{\text{miles}}}{1} \times \frac{1 \text{ km}}{0.62 \ \cancel{\text{miles}}} \approx 48 \text{ km}$$

Always arrange your conversion factors so that you cancel out the part of the unit you want to swap out for something else. Each unit that you don't want in your final answer has to appear in both a numerator and a denominator.

Sometimes you can't do a conversion in one step, but you can string together a series of conversion factors. For instance, here's how you can set up a problem to convert 30 miles per hour to meters per second. Notice how I multiply by a series of fractions, making sure that every unit I want to cancel out appears in the numerator of one fraction and the denominator of another.

$$\frac{30 \text{ mi.}}{1 \text{ hour}} \times \frac{1 \text{ km}}{0.62 \text{ mi.}} \times \frac{1{,}000 \text{ m}}{1 \text{ km}} \times \frac{1 \text{ hour}}{60 \text{ min.}} \times \frac{1 \text{ min.}}{60 \text{ s}} \approx 13 \text{ m/s}$$

Doing speedy metric conversions

In the metric system, one unit can be used as a basis for a broad range of units by adding a prefix (Table 2-2 shows some of the most common prefixes). Each prefix multiplies the base unit by a power of 10. For example, *kilo-* says that the unit is 1,000 times (10^3 times) larger than the base unit, so a kilometer is 1,000 meters. And *milli-* means the unit is 0.001 times (10^{-3}) smaller than the base unit. This means that converting from one metric unit to another is usually a matter of moving the decimal point.

Table 2-2		Metric Prefixes	
Prefix	*Symbol*	*Meaning (Decimal)*	*Meaning (Power of Ten)*
Nano-	n	0.000000001	10^{-9}
Micro-	μ	0.000001	10^{-6}
Milli-	m	0.001	10^{-3}
Centi-	c	0.01	10^{-2}
Kilo-	k	1,000	10^3

By finding the difference in exponents on the power of 10 of your original units and the units you want to convert to, you can figure out how many places to move the decimal point.

For instance, say you have a distance of 20.0 millimeters, and you'd prefer to express it in centimeters. You know that 1 millimeter is 10^{-3} meters, and 1 centimeter is 10^{-2} meters (as Table 2-2 shows). If you find the difference in exponents, you see that $-3 - (-2) = -1$. The answer is negative, so you just have to move the decimal point one place to the left (for a positive answer, you move it to the right). Thus, 20.0 millimeters is equal to 2.00 centimeters.

Using temperature-conversion equations

You can use the following equations to convert between the different units of temperature:

- ✔ Kelvin temperature = Celsius temperature + 273.15
- ✔ Celsius temperature = ⅝(Fahrenheit temperature − 32°)

Keeping it short with scientific notation

Physicists often delve into the realms of the very small and the very large. Fortunately, they also have a very neat way of writing very large and very small numbers: *Scientific notation*. Essentially, you write each number as a decimal (with only one digit to the left of the decimal point) multiplied by 10 raised to a power.

Say you want to write down the speed of light in a vacuum, which is about three hundred million meters per second. This is a three followed by eight zeros, but you can write it as just a 3.0 multiplied by 10^8:

$$300,000,000 \text{ m/s} = 3.0 \times 10^8 \text{ m/s}$$

You can write small numbers by using a negative power to shift the decimal point to the left. So if you have a distance of 4.2 billionths of a meter, you could write it as

$$0.0000000042 \text{ m} = 4.2 \times 10^{-9} \text{ m}$$

Note how the 10^{-9} moves the decimal point of the 4.2 nine places to the left.

Brushing up on basic algebra

To do physics, you need to know basic algebra. You're going to be slinging some equations around, so you should be able to work with variables and move them from one side of an equation to the other as needed, no problem.

You don't need to be bogged down or daunted by the formulas in physics — they're only there to help describe what's going on in the real world. When you see a new formula, consider how the different parts of the equation relate to the physical situation it describes.

Take a simple example — the equation for the speed, v, of an object that covers a distance Δx in a time Δt (**Note:** The symbol Δ means "change in"):

$$v = \frac{\Delta x}{\Delta t}$$

Before you go any further, try relating the parts of this equation to what you intuitively understand about speed. You can see in the equation that if Δx increases, then v increases — if you cover a greater distance in a given time, then you're traveling faster. You can also see that if Δt (in the denominator of the fraction) increases, then v decreases — if it takes you longer to cover a given distance, then you're moving more slowly.

If you need to, you can rearrange an equation algebraically to isolate the part you're interested in. That way, you can get a feel for how other variables affect each other. For instance, see what the equation means for travel time by rearranging it to isolate Δt:

$$\Delta t = \frac{\Delta x}{v}$$

Now you can see that Δt increases as Δx increases, and Δt decreases as v increases. This just means that travel time increases if you have to travel farther and decreases if you travel faster.

Using some trig

You work with some angles to this book — such as those you have to figure out when light bounces off mirrors or bends in lenses. To handle angles and related distances, you need some trigonometry.

Pretty much everything in trig comes down to the right triangle. For example, take a look at the right triangle in Figure 2-1. The two shorter sides, or legs, are called x and y (because they lie along the x- and y-axes respectively), and the longest side, across from the 90° angle, is the *hypotenuse*. One of the other internal angles is marked θ.

Figure 2-1:
The two legs
(x and y)
and hypot-
enuse (h)
of a right
triangle.

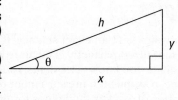

Here's one important formula to know: the Pythagorean theorem. It relates the lengths of x, y, and h, so given the lengths of two sides, you can find the length of the third:

$$x^2 + y^2 = h^2$$

To work with angles (such as θ), you need the trig functions sine, cosine, and tangent. To find the values of trig functions, just divide one side of the triangle by another. Here's what it looks like:

- ✔ **Sine:** $\sin\theta = \dfrac{y}{h} = \dfrac{\text{opposite side}}{\text{hypotenuse}}$

- ✔ **Cosine:** $\cos\theta = \dfrac{x}{h} = \dfrac{\text{adjacent side}}{\text{hypotenuse}}$

- ✔ **Tangent:** $\tan\theta = \dfrac{y}{x} = \dfrac{\text{opposite side}}{\text{adjacent side}}$

Note that these equations relate any two sides of a right triangle to the angle that's between the hypotenuse and one of the other sides. So if you know θ and one of the other sides, you can use some algebra (and your calculator) to find the length of any other side.

To find the angle θ, you can go backward with inverse sines, cosines, and tangents, which are written like this: $\sin^{-1}$, $\cos^{-1}$, and $\tan^{-1}$. If you make the appropriate fraction out of two known sides of a triangle and take the inverse sine of that ($\sin^{-1}$ on your calculator), it'll give you back the angle itself. Here's how the inverse trig functions work (see Figure 2-1 for which sides are x, y, and h):

- ✔ **Inverse sine:** $\sin^{-1}\left(\dfrac{y}{h}\right) = \theta$

- ✔ **Inverse cosine:** $\cos^{-1}\left(\dfrac{x}{h}\right) = \theta$

- ✔ **Inverse tangent:** $\tan^{-1}\left(\dfrac{y}{x}\right) = \theta$

Physicists use sine and cosine functions to describe real-world waves and alternating current and voltage. I introduce waves in Chapter 6, and I cover alternating current (AC circuits) in Chapter 5.

Using significant digits

You may be surprised to hear that physics isn't an exact science! It can be pretty accurate, but nothing is ever measured perfectly. The more accurately the quantity is measured, the more digits you know. The digits you know are the *significant figures*. For instance, a stopwatch measurement of 11.26 seconds has four significant figures. Here are a few guidelines for figuring out what's significant:

- For a decimal less than 1, everything that follows the first nonzero digit is significant. For example, 0.0040 has two significant digits.

- For a decimal greater than 1, all digits, including zeros after the decimal point, are significant. For instance, 20.10 has four significant digits.

- For a whole number, the non-zero digits are significant. Any number of trailing zeros also may be significant.

So how do you show the accuracy of a measurement such as 1,000 meters, which ends in zeros? You may know anywhere between one and four digits from your measurements. The best way to clear this up is to use scientific notation. For instance, if you write 1,000 as 1.000×10^3, with three zeroes after the decimal point, the number has four significant figures — you measured to the nearest meter. If you write it as 1.00×10^3, with two zeroes after the decimal point, the number has three significant figures — you measured to the nearest ten meters. (For info on scientific notation, see the earlier section "Keeping it short with scientific notation.")

When you do calculations with numbers that are known only to a particular accuracy, then your answer is also only of a particular accuracy. After you do all your calculations, you need to round the answer. Here are some simple rules you can apply:

- **If you multiply or divide two numbers:** The answer has the same number of *significant figures* as the least-accurate of the two numbers being multiplied or divided. For example, consider the following calculation:

    $12.45 \times 0.050 = 0.6225$

    Because 0.050 has two significant figures, you round the answer to 0.62.

- **If you add or subtract two numbers:** The answer has the same number of *decimal places* as the least-accurate of the two numbers you're adding or subtracting. For example, consider

    $11.432 + 1.3 = 12.732$

    Because the least-accurate number, 1.3, has only one decimal place, write the answer as 12.7.

Refreshing Your Physics Memory

To make progress, physics often builds on previous physics advances. For example, knowing about vectors is important not just to handle problems with acceleration (that's Physics I) but also to help you track charged particles in magnetic fields (that's Physics II).

In this section, you take a down-memory-lane tour of some Physics I concepts that pop up again in Physics II. If you don't feel comfortable with these topics, check out a physics text to make sure you're up to speed in Physics I before proceeding.

Pointing the way with vectors

Vectors are the physics way of pointing a direction. A *vector* has a direction and a magnitude (size) associated with it — the *magnitude* is the vector's length.

You usually see the names of vectors in bold in physics. Figure 2-2 shows vector **A.** That's just a standard vector, and it may stand for, say, the direction an electron is traveling in. The length of the vector may indicate the speed of the electron — the faster the electron is going, the longer the vector.

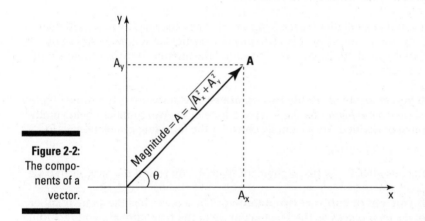

Figure 2-2:
The compo-
nents of a
vector.

You don't see lots of vectors in this book (did I just hear a sigh of relief?), but you should know how to break a vector like **A** up into its components along the *x*- and *y*-axes (you need to do this in Chapter 4 for the magnetic field and in Chapter 5 for alternating currents and voltages).

If you're given the length of the vector (its magnitude, labeled A in Figure 2-2) and the angle θ (its direction), breaking a vector into its components works like this:

$\blacktriangleright$ $A_x = A \cos \theta$

$\blacktriangleright$ $A_y = A \sin \theta$

where A_x is the x component of vector A and A_y is the vector's y component. (This is really just a bit of trig, where A_x and A_y are the legs of the triangle and A is the hypotenuse — see the earlier section "Using some trig" for info on the sine and cosine functions.)

Resolving vectors into components is particularly useful if you have to add two vectors, $A + B$. You break them up into their separate components and then add those components to get the components of the vector sum, which is a new vector you can call C:

✔ $C_x = A_x + B_x$

✔ $C_y = A_y + B_y$

When you have the components of a vector like C, you can covert them into a length (magnitude) for C (written as $|C|$) and an angle for C this way:

✔ **Magnitude of C:** $|C| = \sqrt{C_x^2 + C_y^2}$

   Note: This is just the Pythagorean theorem solved for the hypotenuse $|C|$.

✔ **Direction of C:** $\tan^{-1}\left(\dfrac{C_y}{C_x}\right) = \theta$

   See the earlier section "Using some trig" for info on inverse trig functions.

So now you're able to go from representing a vector in terms of its length and angle to its components and then back again — a very handy skill to have.

Moving along with velocity and acceleration

This book has a little to say about velocity and acceleration. For example, you work with them when a magnetic field diverts electrically charged particles from the direction in which they're traveling.

Both velocity and acceleration are vectors, v and a respectively. *Velocity* is the change in the position-vector divided by the time that change took. For example, if the position of a ping-pong ball is given by the position-vector x, then the change in the position (Δx) divided by the amount of time that change took (Δt) is the velocity:

$$v = \frac{\Delta x}{\Delta t}$$

As a vector, velocity has a direction. The magnitude of the velocity vector is the *speed*, which has a size but not a direction. That is, velocity is a vector, but speed isn't.

If the velocity isn't staying constant, the ping-pong ball is undergoing *acceleration*. *Acceleration* is defined as the change in velocity divided by the time that change takes, or

$$a = \frac{\Delta v}{\Delta t}$$

Note that a change in direction is considered a change in velocity, so something can be accelerating even if its speed doesn't change.

Velocity is commonly measured in meters per second (m/s) — which means that acceleration's units are commonly meters per second squared (m/s^2).

Strong-arm tactics: Applying some force

When an electron enters an electric field, it gets pushed one way or another — that is, it experiences a *force*. Physics I has a lot to say about force — for example, here's the famous equation that relates total force *(F)*, mass *(m)*, and acceleration *(a)* (note that acceleration and force are both vectors):

$$F = ma$$

So to find out how much force is acting on the electron to push it along (and you don't need much, because electrons don't weigh very much), you'd put in the electron's acceleration and its mass, and you'd get the total force acting on it. The formula also shows that applying a force to something can make it accelerate, and you see that idea used every now and then in this book.

The units of force you see most commonly are *newtons* (in the meter-kilogram-second system), symbol N, named for Sir Isaac Newton (the fellow with the falling apple acted on by the force of gravity).

Getting around to circular motion

Charged particles in magnetic fields travel in circles, so you need to know something about circular motion in Physics II. Physics I has plenty to say about circular motion. For example, take a look at Figure 2-3, where an object is traveling in circular motion.

The velocity of an object moving in a circle points along the circle of its path — this is called the *tangential* direction. The force that keeps the object moving in a circle points toward the center of the circle — in a direction that's at right angles to the velocity. For instance, when you spin a ball on a string, the string can exert a force on the ball only in the direction that's along its length, perpendicular to the path of the ball; this is what causes the ball to move in a circular path.

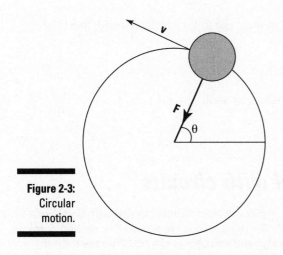

Figure 2-3:
Circular
motion.

The angle that an object moving in circular motion covers in so many seconds is its *angular velocity,* ω:

$$\omega = \frac{\Delta\theta}{\Delta t}$$

Here, the angle θ is measured in radians, so the units of angular velocity are radians/second. (***Note:*** Exactly 2π radians are in a complete circle, which means that 2π radians equals 360°, or each radian is 360° ÷ 2π degrees.)

If the object is speeding up or slowing down, it's undergoing *angular acceleration,* which is given the symbol α. *Angular acceleration* is defined as the change in angular velocity (Δω) divided by the time that change took (Δ*t*):

$$\alpha = \frac{\Delta\omega}{\Delta t}$$

The units of angular acceleration are radians/second^2.

In circular terms, force becomes *torque,* with the symbol **τ** (also a vector, of course), where the magnitude of torque equals force multiplied by distance and the sine of the angle between them:

$$\tau = Fr \sin\theta$$

And the counterpart of mass in circular terms is the *moment of inertia, I.* Newton's law, force = mass × acceleration, becomes this in circular terms:

$$\tau = I\alpha$$

That is, torque = moment of inertia × angular acceleration.

Even linear kinetic energy has an alter ego in the circular world, like this:

$$KE = \frac{I\omega^2}{2}$$

You can have angular momentum, *L,* as well:

$$L = I\omega$$

Getting electrical with circuits

Physics I introduces the idea of circuits, at least simple circuits with batteries. The rules of resistance and Kirchoff's rules, which I review in this section, form the basis for describing the currents and voltages in circuits. You need these rules whenever you work out the various currents and voltages. For example, in Chapter 5, you use them for a simple circuit with three elements in series. You can find a more thorough description of these rules in *Physics For Dummies*.

According to Ohm's law, you can determine the current going through any resistor with the following equation, where *I* is the current measured in amperes, *V* is the voltage across the resistor measured in volts, and *R* is the resistance of the resistor measured in ohms (Ω):

$$I = \frac{V}{R}$$

That helps with individual resistors, but what about when they're assembled into a circuit as Figure 2-4 shows? There, you can see three resistors with resistances of 2 Ω, 4 Ω, and 6 Ω. The currents in each wire, I_1, I_2, and I_3, are driven by the two batteries, which generate voltages of 12 volts and 6 volts.

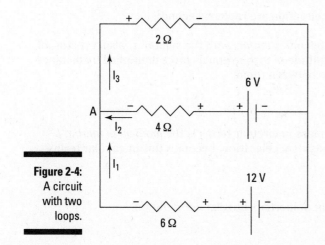

Figure 2-4:
A circuit
with two
loops.

To solve for the currents and voltages, you use Kirchoff's rules:

- ✔ **The loop rule:** The sum of voltages (ΣV) around a loop — any loop in the circuit — is zero:

    $\Sigma V = 0$ around a loop

- ✔ **The junction rule:** The sum of all currents (ΣI) into any point in the circuit must equal the sum of all currents out of that point (that is, the net sum of all currents into and out of any point in the circuit must be zero):

    $\Sigma I = 0$ at any point in the circuit

Part II
Doing Some Field Work: Electricity and Magnetism

The 5th Wave   By Rich Tennant

In this part . . .

Physicists have long been friends with electricity and magnetism. In this part, you see all about electric field, charges, the force between charges, electric potential, and more. You also explore magnetism, such as the magnetic field from a wire, the force between two wires, how charged particles orbit in magnetic fields, and the like. And you look at AC circuits, in which magnetic and electric fields work together to regenerate the current.

Chapter 3

Getting All Charged Up with Electricity

*T*his chapter is dedicated to things that go *zap*. Chances are that your day-to-day life would be very different without electrical appliances from the computer to the light bulb. But electricity is even more important than that; it's a physical interaction that's fundamental to how the whole universe works. For example, chemical reactions are all basically of an electrical nature, and without electricity, atomic matter — the world as you know it — could not exist.

Even though electricity is integral to the existence of all the complicated constructions of matter and chemistry, it has a simple and beautiful nature. In this chapter, you see what makes static electricity, electric fields, and electric potential work.

Understanding Electric Charges

Where does electric charge come from? It turns out to be built into all matter. An atom is made up of a nucleus and electrons in orbit around the nucleus, and the nucleus is made up of protons and neutrons. The protons have a positive charge (+), and the electrons a negative charge (–). So you have electric charges inside any piece of matter you care to name.

What can you do with that charge? You can separate charges from each other and so charge up objects with an excess of one charge or another. Those separated charges exert forces on each other. In this section, I discuss all these concepts — and more — about electric charges.

Can't lose it: Charge is conserved

Here's an important fact about charge: Just as you can't destroy or create matter, you can't create or destroy charge. You have to work with the charge you have.

Because charge can't be created or destroyed, physicists say that charge is *conserved*. That is, if you have an isolated system (that is, no charge moves in or out of the system), the net charge of the system stays constant.

Notice that the conservation of charge says the *net* charge remains constant — that is, the sum of all charges stays the same. The actual distribution of charges can change, such as when one corner of a system becomes strongly negatively charged and another corner becomes positively charged. But the sum total of all charge — whether you're talking about the whole universe or a smaller system — remains the same. No charge in or out means the charge of the system remains constant.

Measuring electric charges

Electric charges are measured in *coulombs* (C) in the MKS system, and each electron's charge — or each proton's charge — is a tiny amount of coulombs. The proton's charge is exactly as positive as the electron's charge is negative.

The electric charge of a proton is named *e,* and the electric charge of an electron is –*e.* How big is *e?* It turns out that

$$e = 1.60 \times 10^{-19} \, \text{C}$$

That's really tiny. Here's how many electrons make up 1 coulomb:

$$\frac{1 \, \text{coulomb}}{1.60 \times 10^{-19} \, \text{coulombs/electron}} = 6.25 \times 10^{18} \, \text{electrons}$$

So there are 6.25×10^{18} electrons in 1 coulomb of (negative) charge.

Opposites attract: Repelling and attracting forces

An uncharged atom is composed of as many electrons as protons. These stay together because of the mutual attraction of these positively and negatively charged components. That's why all ordinary matter doesn't instantly disintegrate before your very eyes.

Objects that have the same charge (– and – or + and +) exert a repelling force on each other, and objects that have opposite charges (+ and –) attract each other (you've always heard that opposites attract, right?). For example, take a look at Figure 3-1, where two suspended ping-pong balls are being charged in various ways.

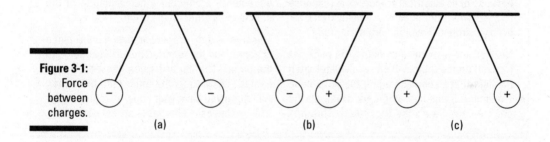

Figure 3-1: Force between charges.

(a)   (b)   (c)

In Figure 3-1a, the two ping-pong balls have the same, negative charge. They're exerting a repelling force on each other, forcing them apart. And the funny thing is that they'll stay that way indefinitely, no additional action or batteries needed — the static charge on each ping-pong ball just stays there.

Actually, it's not totally true that the charge on each ping-pong ball just stays there. In fact, charge is continually being transferred from charged objects to the water molecules in the air, which carry it off. On humid days, charged objects actually lose their charge faster.

In Figure 3-1b, the two ping-pong balls have opposite charges, – and +, so they attract each other. Note that if they were to touch, charge would flow and the two ping-pong balls would each end up with the same charge. If the + charge is the same magnitude as the – charge, that means that the balls would end up electrically neutral and would just hang down straight.

In Figure 3-1c, the two ping-pong balls have the same positive charge, so once again they repel each other. The force repelling the two ping-pong balls is the same as that in Figure 3-1a if the positive and negative charges have the same magnitude (just different signs).

TECHNICAL STUFF

Putting charges to work: The story on photocopiers

Many areas of modern life depend on electrical charges — and not just flowing electricity. Static electricity plays a role as well. For example, take a photocopier, which makes copies through a process called *xerography* (from the Greek words *xeros* and *graphos*, which mean "dry writing").

Here's how a photocopier works: A drum with a surface that contains the element selenium (Se) does the actual printing — selenium is used because of its electrical properties in response to light. The drum is given a positive charge, evenly distributed over the drum's surface.

Next, an image of the document to be copied is focused on the drum, which revolves to catch the light that's scanning along the document. The image forms light and dark areas on the drum — and here's the tricky part: The dark parts retain their positive charge, but thanks to the properties of selenium, the light areas become conducting, and their positive charge is conducted away, leaving them neutral.

In the next step, dry toner (powder of the color you're printing — black powder in a black-and-white copier) is given a negative charge and sprayed onto the drum. The negatively charged toner sticks to the positively charged areas of the drum (which mimic the dark areas of the document), producing a mirror image of the document in toner on the drum.

The drum is then pressed against a blank piece of paper, and the toner on the drum adheres to the paper. The paper is then passed through heated rollers to fix the toner onto the paper. And there you have your copy, all thanks to the ability of opposite charges to attract each other.

Getting All Charged Up

In this section, you see how to deliver charge to objects. I let it sit there a while, and you experience charge in a way that really makes your hair stand on end: static electricity. I also transfer charge from object to object, and I let it flow nice and smoothly through wires.

Static electricity: Building up excess charge

You may not be able to create or destroy charges, but you can move them around and create imbalances within a system. When you charge up an object, you keep adding more and more charges to the object, and if that charge has nowhere to go, it just accumulates. *Static electricity* is the kind of electricity that comes from this excess charge.

Everyone's familiar with the unpleasant experience of walking across a rug and then getting zapped by a doorknob. What's actually happening there? Turns out that you're actually picking up spare electrons from the rug. Your body collects an excess of electrons, and when you touch the doorknob, they flow out of you. Ouch!

Protons, being bound inside the nucleus, aren't really free to flow through matter, so when you charge something, it's usually the electrons that are moving around and redistributing themselves. When something gets charged negatively, electrons are added to it. When it gets charged positively, electrons are taken away, leaving the protons where they are, and the net surplus of protons makes a positive charge.

Before you're zapped, that excess of charge is static electricity: It's *electricity* because it's made up of electrical charges, and it's called *static electricity* when it isn't flowing anywhere. When you're charged with static electricity, each of the hairs on your head carries a share of this excess of electrons. You may develop a spiky new hairstyle as each hair repels its neighbor (which has the same charge). Your hair quickly returns to normal if you touch something that the excess electrons can flow into. They quickly rush through the contact point, giving you the shock.

Though charge can flow through your fingertip, you usually find it flowing through wires in a circuit, where it doesn't build up. In circuits, charge doesn't collect and remain stationary, because it's always free to flow (however, I show you an exception to this idea in the later section "Storing Charge: Capacitors and Dielectrics").

But when electricity gets blocked and yet still piles up — it can't go anywhere — then you have static electricity. If the electricity in a circuit is like a river of electricity that keeps flowing around and around (kept in motion by, say, a battery), then static electricity is like a river of electricity that's dammed up — but charges keep getting added. So although charges don't build up in circuits, they do build up when you have static electricity.

Checking out charging methods

In this section, I cover two ways to charge objects: by contact and by induction. These are simple physical mechanisms that can help you understand how charge behaves and how it can be redistributed.

Charging by contact

Charging by contact is the simplest way of charging objects — you just touch the object with something charged and *zap!* The object becomes charged. No big mystery here.

For example, take a look at Figure 3-2 — a negatively charged rod is brought into contact with a ball that's originally neutral. The result? The ball is left with a negative charge. That's because the electrons in the rod are always pushing each other (because like charges repel), so they're always looking for ways to redistribute themselves farther apart. When the rod comes in contact with the ball, some of the electrons take the opportunity to slip off the rod and onto the ball. Presto! The ball gets charged. (*Note:* For this to happen, the electrons need to be free to flow through the materials, which can happen if the materials are *conductors*. Materials that don't allow electrons to flow through them are *insulators*. I discuss both types of materials later in "Considering the medium: Conductors and insulators.")

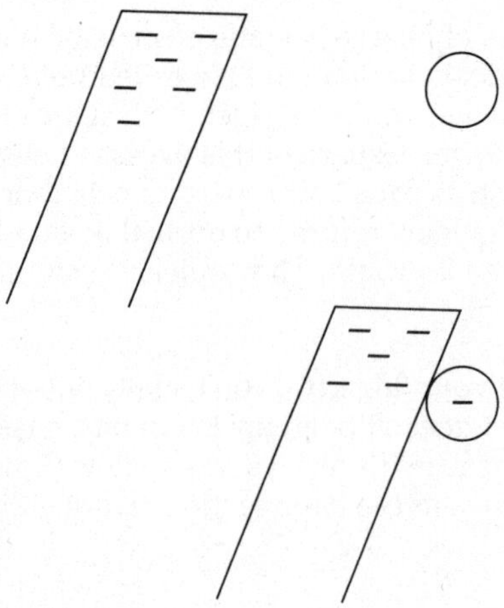

Figure 3-2:
Charging by
contact.

Touch a negatively charged glass rod to a neutral ping-pong ball, and the ball acquires a negative charge by contact. But you may wonder how to charge the glass rod in the first place. You can do this in many ways, but the simplest and oldest way is to take a glass rod and some silk and rub the two together. A transfer of electrons from one material to the other occurs due to molecular forces between the two types of material. Different materials have different propensities to exchange electrons — you may have noticed that a balloon and a wool sweater work well.

Charging by induction

You can deliver charge to an object indirectly using induction. Here's how charging by *induction* works: You bring a charged rod close to a neutral object. Say the rod is charged negatively — the negative charges (electrons) in the neutral object are repelled to the opposite side of the object, leaving a net positive charge close to the rod, as Figure 3-3 shows at the top.

Now comes the clever part: You connect the far side of the object to the ground. Just connect a wire from the far side of the object to the actual Earth, which acts as a huge reservoir of charge. The negative charges — the electrons — that are being forced to the far side of the object are frantic to get off the object, because the charge on the rod is repelling them. By connecting the far side of the object to the ground with a wire, you provide those electrons with an escape route. And the electrons take that escape route by the millions and trillions.

Then you cleverly disconnect the wire from the far side of the object. The electrons that wanted to get away have fled — and now there's nowhere else for any other charges to go. The result is that the object is left with a positive charge, because you've drained off much of the negative charge that was being pushed by the rod. And you haven't lost any of the charge on the glass rod.

The upshot is that the object is left with the opposite charge of the rod. And that's charging by induction. Pretty cool, eh?

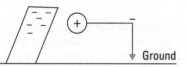

Figure 3-3: Charging by induction.

Lightning rods work through induction. In thunderclouds, charges get separated from the top to the bottom of the cloud, so the top and the bottom of the cloud become strongly charged. When lightning strikes, the charge on the bottom of the cloud is zapping the Earth. If you have a lightning rod, the strong charge on the bottom of the cloud induces the opposite charge on the lightning rod (which is connected to the ground). When lightning strikes, it's attracted to that opposite charge and hits the lightning rod.

Considering the medium: Conductors and insulators

You're probably familiar with the concepts of electrical conductors — like the copper wire in an extension cord — and electrical insulators — like the plastic that coats the electrical wire and prevents the electricity in the wire from delivering a nasty shock. In this section, you take a closer look at conductors and insulators in physics terms.

Say you have two charged objects some distance apart from each other. They're not losing charge; they're just sitting there. Then you bring a piece of rubber and touch both of them with the rubber. What happens? Nothing, because rubber is an electrical *insulator;* electricity is conducted through rubber only with difficulty.

Now say that you bring a copper bar in contact with the two objects — immediately, charge flows from one to the other, because copper is an electrical conductor.

Good electrical *conductors* consist of atoms for which the outermost electrons are not very tightly bound, so they can easily hop from atom to atom and take part in an electric current. The electrons in the very outer orbit around the nucleus are called *valence electrons,* and those are the ones that can detach from atoms and roam freely through the conductor. (Interestingly, materials that are good electrical conductors, like most metals, are usually also good thermal conductors.)

Current is always defined as the direction of the flow of positive charges, but in reality, it's the electrons that do the moving and therefore transport electrical charge. In this case, electrons move from the negatively charged object to the positively charged object. But if you want to draw the direction of the current, that goes from the positively charged object to the negatively charged one. Historically, this convention was adopted before people knew that electrons, not positive charges, carry the current.

Coulomb's Law: Calculating the Force between Charges

Coulomb's law is one of the physics biggies. That's the same Coulomb (Charles-Augustin de Coulomb) that the unit of charge, the *coulomb,* is named after, so you know Coulomb's law has to be some serious stuff.

And serious stuff it is: Charges can attract or repel each other, and Coulomb's law lets you calculate the exact force that two point charges a certain distance away will exert on each other. A *point charge* just has all its charge concentrated in a single point, with no surface area that the charge can be distributed over. Point charges are particularly beloved by physicists because they're easy to work with.

Say you have two point charges, with opposite signs, attracting each other from a certain distance r apart. What's the force between the two charges? Coulomb has the answer: His law says that if the charge of one charge is q_1 and the charge of the other charge is q_2, then the force between the two charges is

$$F = \frac{kq_1q_2}{r^2}$$

In this equation, k is a constant, and its value is 8.99×10^9 N-m²/C²; q_1 and q_2 are the charges, in coulombs (C), of the charged objects doing the attracting or repelling; r is the distance between the charges; and F is the electrostatic force between the charges.

Force is a vector, so when you're looking at point charges, the direction of the force is always along a line between the two charges (assuming there are only two) and

✔ Toward each other if the charges have opposite signs (that is, the force has a negative sign)

✔ Away from each other if the charges have opposite signs (that is, the force has a positive sign)

Introducing Electric Fields

Electric field is the field in space created by electric charges. When two charges attract or repel each other, their electric fields are interacting.

Charge can be distributed in many ways. You can have point charges, sheets of charge, cylinders of charge, and many more configurations. Coulomb's law, which I discuss earlier in "Coulomb's Law: Calculating the Force between Charges," works only for point charges. So how do you handle force for other distributions of charge? You often use electric fields, which I discuss in this section.

Sheets of charge: Presenting basic fields

How do you calculate force for a sheet of charge? Instead of modifying Coulomb's law to handle sheets of charge, you can simply measure the force that the sheet of charge exerts on a small positive test charge. From there, you know how much force per coulomb the sheet of charge is capable of exerting, and when you have your own charge, which may be positive or negative, you can simply multiply the force per coulomb by the size of your charge.

The idea of measuring force per coulomb to handle non-point charges got to be very popular and became known as the *electric field*. Here's the definition: *Electric field (E)* is the force *(F)* that a small test charge would feel due to the presence of other charges, divided by the test charge *(q)*:

$$E = \frac{F}{q}$$

Electric field's units are newtons per coulomb (N/C), and electric field is a vector. The direction of the electric field at any point is the force that would be felt by a *positive* test charge.

What's that in plain English? Electric field is just the force per coulomb that a charge would feel at any point in space. You divide out the test charge to leave you just newtons per coulomb, which you can multiply by your own charge to determine the force that charge would feel.

For example, say you're walking on a wool carpet and pick up a static electricity charge of -1.0×10^{-6} coulombs. You suddenly encounter a 5.0×10^6 N/C electric field in the opposite direction in which you're walking, as Figure 3-4 shows.

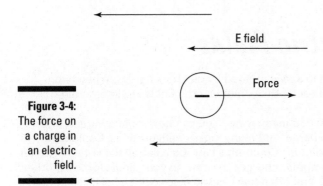

Figure 3-4:
The force on
a charge in
an electric
field.

How big of a force do you feel? Well, the electric field is 5.0×10^6 newtons per coulomb and you have a charge of -1.0×10^{-6} coulombs, so you get the following:

$$F = qE = (-1.0 \times 10^{-6} \text{ C})(5.0 \times 10^6 \text{ N/C}) = -5.0 \text{ N}$$

That is, you feel a force of 5.0 N, and the minus sign means that the force is in the direction opposite of the electric field. That's a little more than 1 pound of force.

So that's what electric field tells you — how much electric influence is in a given region, ready to cause a force on any charge you bring into the electric field.

Note that the electric field has a direction. How can you tell which way the force that an electric field causes will push the charge you bring into the electric field? You can do this the hard way, with formal definitions, or you can use the easy way. I prefer the easy way. Just think of electric field as coming from positive charges — that is, electric field arrows always do the following:

✔ Point away from any positive charges that create the electric field

✔ Go into negative charges

So you can always think of a bunch of positive charges as sitting at the base of the electric field arrows, and that tells you which way the force will act on the charge you bring into the electric field. For example, you have a negative charge in Figure 3-4, and you can think of the electric field arrows as coming from a bunch of positive charges — and because like charges repel, the force on your charge is away from the base of the arrows.

Looking at electric fields from charged objects

Not all electric fields are going to be as polite and evenly spaced as the electric field associated with the sheet of charge you see in Figure 3-4. For example, what's the electric field from a point charge?

Say that you have a point charge Q and a small test charge q. How do you find the force per coulomb? Coulomb's law to the rescue here — just plug in the charges Q and q (for q_1 and q_2) and the distance between them to get the size of the force (see the earlier section "Coulomb's Law: Calculating the Force between Charges" for more on this formula):

$$F = \frac{kQq}{r^2}$$

So what's the electric field? You know that $E = F/q$, so all you have to do is to divide by your test charge, q, to get the following:

$$E = \frac{F}{q} = \frac{kQ}{r^2}$$

So the electric field at a given place falls off by a magnitude of r^2, the square of the distance away from a point charge.

And what about the direction of the electric field? Well, the force exerted by a point charge is radial (that is, toward or away from the point charge). And electric field emanates from positive charges and goes into negative charges, so Figure 3-5 shows you what the electric field looks like for a positive point charge and a negative point charge.

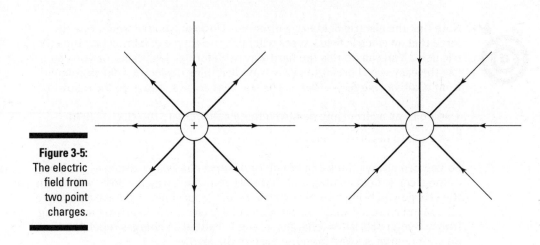

Figure 3-5:
The electric
field from
two point
charges.

Uniform electric fields: Taking it easy with parallel plate capacitors

The electric field between multiple point charges isn't the easiest thing to come to grips with in terms of vectors. So to make life easier, physicists came up with the parallel plate capacitor, which you see in Figure 3-6.

A *parallel plate capacitor* consists of two parallel conducting plates separated by a (usually small) distance. A charge $+q$ is spread evenly over one plate and a charge $-q$ is spread evenly over the other. That's great for physicists' purposes, because the electric field from all the point charges on these plates cancels out all components except the ones pointing between the plates, as you see in Figure 3-6.

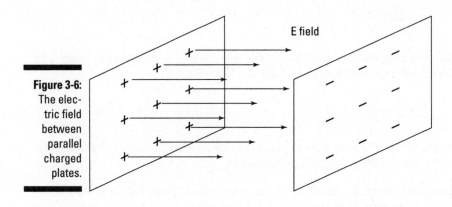

Figure 3-6:
The elec-
tric field
between
parallel
charged
plates.

E field

So by being clever, physicists arrange to get a constant electric field, all in the same direction, which is a heck of a lot easier to work with than the field from point charges.

So what is the electric field between the plates? You can determine that the electric field (E) between the plates is constant (as long as the plates are close enough together), and in magnitude, it's equal to

$$E = \frac{q}{\varepsilon_0 A}$$

where ε_o, the so-called *permittivity of free space,* is 8.854×10^{-12} $C^2/(N\text{-}m^2)$; q is the total charge on either of the plates (one plate has charge $+q$ and the other is $-q$); and A is the area of each plate in square meters.

The equation for the electric field (E) between the plates of a parallel plate capacitor is often written in terms of the *charge density,* σ, on each plate, where $\sigma = q/A$ (the charge per square meter), and here's what that makes the equation look like:

$$E = \frac{q}{\varepsilon_0 A} = \frac{\sigma}{\varepsilon_0}$$

When you work with a parallel plate capacitor, life becomes a little easier because the electric field has a constant value and a constant direction (from the + plate to the − plate), so you don't have to worry about where you are between those plates to find the electric field.

Take a look at an example. Say, for instance, that you put a positive charge of +1.0 coulombs inside the plates of a parallel plate capacitor, as in Figure 3-7.

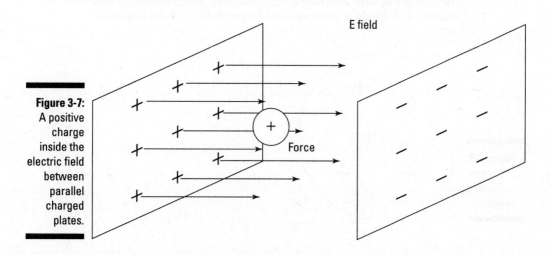

Figure 3-7:
A positive
charge
inside the
electric field
between
parallel
charged
plates.

And also assume that the charge of the plates is 1.77×10^{-11} coulombs and that the area of each plate is 1.0 square meters. That would give you the following result for the electric field between the plates:

$$E = \frac{1.77 \times 10^{-11} \text{C}}{\left(8.854 \times 10^{-12} \text{C}^2/\text{N} \cdot \text{m}^2\right)\left(1.0 \text{ m}^2\right)} \approx 2.0 \text{ N/C}$$

The electric field is a constant 2.0 newtons per coulomb. To find the force on a 1.0-coulomb charge, you know that

$$F = qE$$

And in this case, that's 1.0 C × 2.0 N/C, for a total of 2.0 N (or about 0.45 pounds). That calculation is pretty simple, because the electric field between the parallel plates is constant — unlike what the electric field would be between two point charges.

Shielding: The electric field inside conductors

This section examines how, in any electric field at all, people can make a little safe haven — a region of zero electric field — with only the aid of a hollow conductor!

Figure 3-8 shows an internal cross-section of a spherical conducting ball. Say that some charges were planted inside the solid metal ball. Now, electric charges always generate an electric field, and conducting materials let charges flow freely in response to electric field, so what happens?

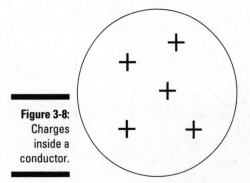

Figure 3-8:
Charges
inside a
conductor.

The electric field generated by the charges implanted in the conducting material push other, similar charges away. As a result, charges move around until like charges are as far away from each other as possible. You can see the result in Figure 3-9 — charges immediately appear on the surface of the conductor. In fact, all the charge appears immediately on the surface — no net charge is left inside the conductor. (*Note:* Although this example shows positive charge moving, electrons really do the moving, so a reduction of electrons that appear on the surface is what creates a positive charge.)

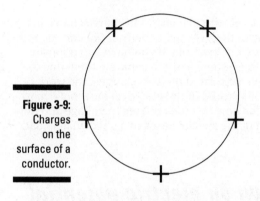

Figure 3-9:
Charges
on the
surface of a
conductor.

This kind of behavior — the free motion of charges in conductors — is very useful. For example, if you're in the middle of a region of electric field and you want to have no electric field present, you can shield yourself from the electric field.

To *shield* yourself from the electric field, you place a conducting container in the field, as in Figure 3-10. The electric field outside the container induces a charge on the container. But the nature of conductors is to let charges flow if there's any net charge — and any way for it to move around — so the electric field from the induced charges and the preexisting electric field cancel each other out. The result is that there's no electric field inside the conducting container. You've shielded yourself from the external electric field. Nice.

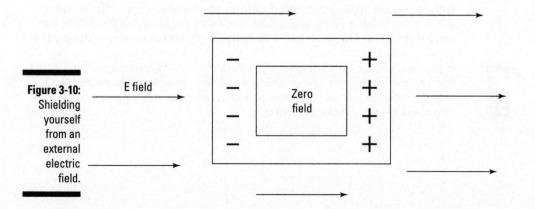

Figure 3-10:
Shielding
yourself
from an
external
electric
field.

Voltage: Realizing Potential

This section discusses an electrical concept that's surely familiar: the idea of *voltage*. Yes, the kind of voltage you get when you plug something into a wall socket. The kind you get when you put a battery into a flashlight or when you rely on a car battery to start your car.

Here, you see how voltage relates to electrical energy. As charges move in electric fields, they can swap some of their energy of motion *(kinetic energy)* for electrical energy, and vice versa. For example, if you have two opposite charges close together, you have to do some work to separate them due to their mutual attraction. After you've separated the two charges, the work you did doesn't just disappear — it's stored in the electrical *potential energy* between them. This section explains how this idea relates to voltage and how voltage relates to the electric field in the special cases of a uniform field and the field around a point charge.

Getting the lowdown on electric potential

If you have a mass in a gravitational field, it has potential energy. As you throw a ball upward, for example, the kinetic energy of its motion is converted to gravitational potential energy as it reaches its peak, and then the potential energy changes back to kinetic energy as the ball falls back to you. The gravitational forces on the ball do work and exchange potential and kinetic energy. Because a force likewise acts on charges in an electric field, you can speak about potential energy here, too. That potential energy is *electric potential energy*.

What makes all forms of energy essentially the same is that they can all be converted to mechanical work. As you may remember from Physics I, work done *(W)* is the result of a force *(F)* moving a body through a distance *(s)*, and they're related thus: $W = Fs$. The energy in the interaction of a charge with an electric field is converted to work when the electrical forces move the charge. Moving twice as much charge takes twice as much work for the electrical forces. The work that's done for every unit of charge is the *voltage*.

In physics, voltage is called *electric potential* (not *electric potential energy,* which isn't per unit of charge); sometimes, it's just shortened to *potential*. Instead of using the term *voltage,* it's more correct to say that electric potential is measured in volts, whose symbol is *V.*

Looking at lightning volts

In a thunderstorm, the clouds are at a different electric potential from the ground. The electric potential becomes too great for the air and the air breaks down, conducting electric charge, so every now and then, lightning strikes between the Earth and the clouds.

How many volts are between clouds and the Earth in a thunderstorm? Plenty. It takes 11,000 volts to make a spark across 1 centimeter of air — and there are 100,000 centimeters in a kilometer (about 0.62 miles), the typical height of a cloud during a thunderstorm. You do the math.

Okay, I'll do the math: that's 11,000 volts/centimeter × 100,000 centimeters = 1.1×10^9 volts — which is indeed plenty of volts when compared to, say, a wall socket that has 110 volts.

In the case of a gravitational field, the gravitational force moves a mass in the direction of lower potential — things fall to the ground because they have lower gravitational potential there. In the same way, electrical forces move charges in the direction of lower electrical potential. The faster that the potential energy drops in that direction, the greater the force.

Now remember that electric field is force per unit charge, and electric potential is potential energy per unit charge. Therefore, the electric field is directed down the gradient (slope) of the electrical potential and has a strength proportional to the steepness of the slope.

The electric potential (*V*) at a particular point in space is the electric potential energy of a test charge located at the point of interest divided by the magnitude of that test charge, like this:

$$V = \frac{PE}{q}$$

So you can think of electric potential as electric potential energy per coulomb.

So by how many volts does one plate of a charged parallel plate capacitor differ from the other plate? It differs by the energy needed to move 1 coulomb of charge from one plate to the other. (Note that volts are the same as joules/coulomb.)

Finding the work to move charges

Say that you're sitting around, dismantling the smoke alarm in your apartment (which won't make your landlord very happy) and you find a 9.0-volt battery.

Whipping out the voltmeter you always carry, you measure the voltage between the terminals as exactly 9.0 volts. Hmm, you think. How much energy does it take to move one electron between the two terminals of the battery — a difference in electrical potential between the terminals of 9.0 volts?

Well, you realize that 9.0 volts is the change in potential energy per coulomb between the terminals. And change in potential energy is equal to work. So how much work does it take to move one electron between the terminals? You start by noting that the electric potential is

$$\Delta V = \frac{W}{q}$$

which means that

$$W = q\Delta V$$

Here, W is the work needed to move charge q across potential difference ΔV.

Now plug in some numbers. The charge of an electron is a miniscule -1.6×10^{-19} coulombs, and the potential difference between the negative and positive battery terminals is 9.0 volts, so

$$W = q\Delta V = (-1.6 \times 10^{-19})(9.0) = -1.4 \times 10^{-18} \text{ J}$$

Therefore, -1.4×10^{-18} joules of work is done as the electron moves between the two terminals of a 9-volt battery.

You may remember the meaning of negative work from Physics I. The work is negative because the electron's potential energy falls — that is, the electric force does the work of moving the electron. Moving the electron in the other direction would require an equal quantity of positive work from you, because you'd have to move it up the potential difference, against the electric field.

Finding the electric potential from charges

Say you have a point charge Q. What's the electric potential due to Q at some distance r from the charge? You know that the size of the force on a test charge q due to the point charge is equal to the following (see the earlier section "Looking at electric fields from charged objects" for details):

$$F = \frac{kQq}{r^2}$$

where k is a constant equal to 8.99×10^9 N-m^2/C^2, Q is the point charge measured in coulombs, q is the charge of the test charge, and r is the distance between the point charge and the test charge.

You also know that the electric field at any point around a point charge Q is equal to the following:

$$E = \frac{kQ}{r^2}$$

Thus, close to the point (when r is small), E is large and the field is therefore strong. As you move away from the point charge, r increases and the electric field quickly becomes weak.

Suppose you place a small test charge, q, in this field and try moving it around. The test charge feels a strong force close to the point charge, which quickly falls away as you move it to a greater distance. If the test charge is of the opposite sign to the point charge, you have to do work to pull it away from the point charge. This means that the test charge has a lower potential energy closer to the point charge (this is reversed if the charges are the same sign).

So what's the electric potential from the point charge? At an infinite distance from the point charge, you can't see it or be affected by it, so set the potential from the point charge to be zero there. As you bring a test charge closer, to a point r away from the point charge, you have to add up all the work you do and then divide by the size of the test charge. And the result after you do turns out to be gratifyingly simple. Here's what that it looks like:

$$V = \frac{kQ}{r}$$

So the electrical potential is large close to the point charge (when r is small) and falls away at greater distances. This idea applies to all point charges, so what does it mean for the electrons orbiting the protons in an atom? How hard do you have to work to pull an electron away from an atom?

First find the electrical potential. The size of the charge of the electron and the proton is 1.6×10^{-19} coulombs. The electron and proton are typically 5.29×10^{-11} meters apart, so the electric potential is

$$V = \frac{kQ}{r} = \frac{\left(8.99 \times 10^{9} \text{ N} \cdot \text{m/C}^2\right)\left(1.60 \times 10^{-19} \text{ C}\right)}{5.29 \times 10^{-11} \text{ m}} \approx 27.2 \text{ volts}$$

That close to the proton, the electric potential is a full 27.2 volts. That's quite something for such a tiny charge!

The amount of energy needed to move an electron through 1 volt is called an *electron-volt* (eV). So you may expect that you'd need 27.2 eV of energy to pull an electron out of the atom. But the electron is moving quickly, so it already has some energy to contribute. In fact, the electron has so much kinetic energy that you need only half that, 13.6 eV of energy, to win the electron from the atom!

Illustrating equipotential surfaces for point charges and plates

To illustrate electric potential, you can draw *equipotential surfaces;* that is, surfaces that have the same potential at every point. Drawing an equipotential surface gives you an idea of what the electric potential from a charge or charge distribution looks like. For example, on one equipotential surface, the potential could always be 5.0 volts or 10.0 volts.

Because the potential from a point charge depends on the distance you are from the point charge, the equipotential surfaces for a point charge are a set of concentric spheres. You can see what they'd look like in Figure 3-11.

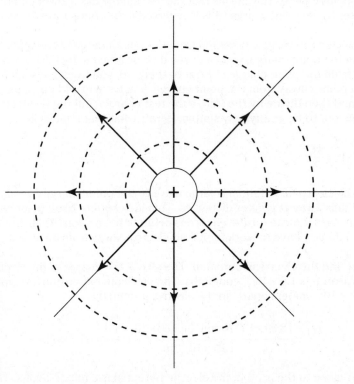

Figure 3-11:
Dashed lines showing equipotential surfaces from a point charge.

Now consider equipotential surfaces between the plates of a parallel plate capacitor (see the earlier section "Uniform electric fields: Taking it easy with parallel plate capacitors" for more on these devices). If you start at the negatively charged plate and move a distance *s* toward the positively charged plate, you know that

$$V = \frac{qs}{\varepsilon_0 A}$$

In other words, the equipotential surfaces here depend only on how far you are between the two plates. You can see this in Figure 3-12, which shows two equipotential surfaces between the plates of the parallel plate capacitor. ***Tip:*** This is analogous to the gravitational potential close to the ground, which increases in proportion to height. They're both cases of uniform fields in which the field is the same at every point.

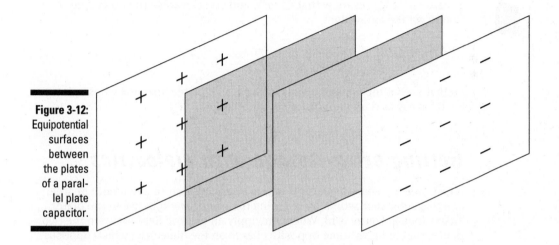

Figure 3-12:
Equipotential
surfaces
between
the plates
of a paral-
lel plate
capacitor.

Storing Charge: Capacitors and Dielectrics

A *capacitor,* generally speaking, is something that stores charge. I discuss parallel plate capacitors earlier in this chapter, but a capacitor need not be shaped like two parallel plates — any two conductors separated by an insulator form a capacitor, regardless of shape. A *dielectric* increases how much charge a capacitor can hold. This section covers how capacitors and dielectrics work together.

Figuring out how much capacitors hold

How much charge is actually stored in a capacitor? That depends on its *capacitance, C.* The amount of charge stored in a capacitor is equal to its capacitance multiplied by the voltage across the capacitor:

$q = CV$

The MKS unit for capacitance C is coulombs per volt (C/V), also called the *farad,* F. For a parallel plate capacitor, the following is true (see the preceding section for details):

$$V = \frac{qs}{\varepsilon_0 A}$$

Because $q = CV$, you know that $C = q/V$, and you can solve the preceding equation to get the following:

$$C = \frac{q}{V} = \frac{\varepsilon_0 A}{s}$$

So that's what the capacitance is for a parallel plate capacitor whose plates each have area A and are distance s apart.

Getting extra storage with dielectrics

A *dielectric* is a semi-insulator that lets a capacitor store even more charge, and this substance works by reducing the electric field between the plates. Take a look at Figure 3-13. When you apply an electric field between the two plates, that induces some opposite charge on the dielectric, which opposes the applied electric field. The net result is that the electric field inside the dielectric (which fills the area between the plates) is reduced, allowing the capacitor to store more charge.

Figure 3-13:
Using a dielectric between the plates of a parallel plate capacitor.

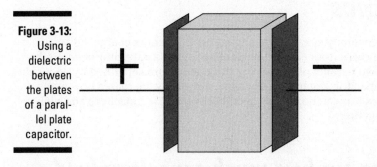

If you fill the space between the plates in a parallel plate capacitor with a dielectric, the capacitance of the parallel plate capacitor becomes the following:

$$C = \frac{\kappa \varepsilon_0 A}{s}$$

The dielectric increases the capacitance of the capacitor by the *dielectric constant,* κ, which differs for every dielectric. For example, the dielectric constant of *mica,* a commonly used mineral, is 5.4, so capacitors that use mica increase their capacitance by a factor of 5.4. The dielectric constant of a vacuum is 1.0.

Calculating the energy of capacitors with dielectrics

Because a capacitor stores charge, it can act as a source of electric current, like a battery. So in terms of capacitance and voltage, how much energy is stored in a capacitor?

Well, when you charge a capacitor, you assemble the final charge q in an average potential V_{avg} (you use the average potential because the potential increases as you add more charge). So here's the energy stored:

$$\text{Energy} = qV_{avg}$$

This raises the question, of course, of just what V_{avg} is. Because the voltage is proportional to the amount of charge on the capacitor (because $q = CV$), V_{avg} is one-half of the final charge. Or looked at another way, as you charge up a capacitor from 0 to its final voltage, the voltage increases linearly, so the average voltage is half the final voltage:

$$V_{avg} = \frac{1}{2}V$$

Plugging this value of V_{avg} into the energy equation and making the substitution that $q = CV$ gives you the following equation:

$$\text{Energy} = \frac{1}{2}CV^2$$

For a parallel plate capacitor with a dielectric, the capacitance is

$$C = \frac{\kappa\varepsilon_0 A}{s}$$

So the energy in a parallel plate capacitor with a dielectric in it is

$$\text{Energy} = \frac{1}{2}\frac{\kappa\varepsilon_0 A}{s}V^2$$

And there you have it — that's the energy stored in a capacitor with a dielectric (the energy ends up in joules from this equation). You can see why dielectrics are considered a good idea when it comes to capacitors — they multiply the capacitance and the energy stored in a capacitor many-fold.

Chapter 4

The Attraction of Magnetism

*L*egend has it that more than 2,000 years ago, Magnes, a Greek shepherd, was walking with his flock when he found the nails that were holding his shoes together became inexplicably stuck to a rock — and that's how magnetism was found. Thousands of years of mystery came together into a scientific understanding only in the last few hundred years.

In this chapter, you explore the physicists' understanding of magnetism. You see why permanent magnets (like the one stuck to Magnes's shoe) attract some apparently nonmagnetic materials (like the iron nail in Magnes's shoe). You see how magnetism is not really a strange new thing but only a different aspect of electricity. And you see how to work out exactly how big a magnetic force is and in which direction it goes. You find the magnetic influence of electrical currents, and you see that electrical currents are the source of magnetism.

With all this knowledge come all kinds of useful devices, like electric motors, speakers, doorbells, and even sophisticated medical imaging machines. That's why I also discuss some practical uses of magnets and magnetism. For instance, I explain how a compass works by moving under the magnetic influence of the molten iron swirling at the center of the Earth. And at last you'll know what makes those things stick to the fridge door.

All About Magnetism: Linking Magnetism and Electricity

Here's the most important thing you need to know about magnetism: It's closely related to electricity (which I cover in Chapter 3). Magnetism and electricity are just different aspects of the same thing — two sides of the same coin. In fact, where you have electricity flowing, you have magnetism, because magnetism comes from electrical current.

In this section, I discuss how current flows even in permanent magnets, because the electrons are in motion in the magnet's atoms. I also explain the repelling and attractive forces at work in magnets, just like the forces between electrical charges. Finally, I give you a formal definition of *magnetic field* that ties magnetism and electric charge together.

Electron loops: Understanding permanent magnets and magnetic materials

Even in bar magnets or refrigerator magnets, electricity is indeed flowing. Every atom in that refrigerator magnet has a bunch of electrons circling a nucleus, and those electrons form an electric current. This is on the level — magnetism in permanent magnets comes from those tiny orbits that electrons are continually zooming around in. Loops of current form a magnetic field, so the atoms behave like tiny magnets.

Most of the time, those tiny magnetic fields are pointing in all different directions, as you see in Figure 4-1a, so they add up to zero. However, in a permanent magnet, those tiny, atomic-level magnets are much more aligned, as in Figure 4-1b. When the micro-magnets in matter align, you end up with magnetism you can measure — and the magnet stays magnetized. That's why you call such magnets (which take no external electrical power) *permanent magnets*. That alignment is the only difference between a permanent magnet and a permanent non-magnet.

Some materials are in between these opposites. For instance, all the atoms, acting like little magnets, may all start out pointing in random directions — imagine that the electrons are orbiting in planes that are randomly orientated. But when you put this material close to a magnet, the atoms are forced to rotate and align all together with the magnet. Then the magnetic influence of each atom adds together, so the material becomes magnetic. But when you take the material away from the magnet, the atoms relax back to their random orientations, and the material stops being magnetic. Such a material is called *paramagnetic*. Two para-magnets won't stick to each other, but they will stick to a permanent magnet.

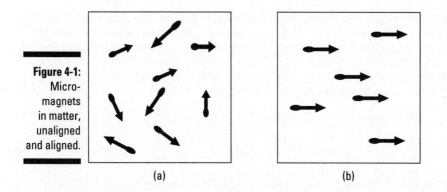

Figure 4-1: Micro-magnets in matter, unaligned and aligned.

(a) (b)

In another type of material, such as iron, the atoms are organized into little aligned regions, called *domains*. Each domain is magnetic, but the material has many domains that are randomly orientated. Now if you place this material near a magnet, the domains are forced to align and the material becomes a magnet. But this time, when you take the material away, the domains stay aligned, and the piece of iron remains magnetic! This type of material is said to be *ferromagnetic*.

Electromagnets are nonpermanent magnets that work only when you have electricity flowing. I discuss these magnets later in "Going to the Source: Getting Magnetic Field from Electric Current."

North to south: Going polar

Electricity has two sides to it: positive and negative. Electric field goes from the plus to the minus side (see Chapter 3 for details). Similarly, magnetism involves *magnetic poles*. And just as electric field goes from + charges to − charges, magnetic field goes from one pole to the other — from *north* to *south*.

The names of the magnetic poles come from the popular use of magnets in compasses — the North and South Poles are used in navigation. The north pole of a permanent magnet automatically points toward magnetic north of the Earth.

Magnetic field is often drawn as a set of lines — that is, magnetic field lines, much as electric field is drawn as electric field lines. Figure 4-2 shows the magnetic field lines going from north pole to south of a permanent magnet.

The magnetic Earth

The Earth is a huge magnet, as anyone with a compass knows — just watch your compass needle point unerringly toward magnetic north. When you change locations, the compass needle finds magnetic north for you again. Unfortunately, magnetic north doesn't match the true North Pole of the Earth — that is, geographic north, where the Earth's axis of rotation pierces the surface of the Earth. The following figure shows how the Earth's geographic north pole is offset from the magnetic north pole.

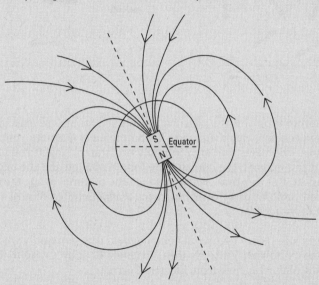

Notice how the magnetic pole in the figure is labeled *S*. That's no typo — opposite magnetic poles attract, so to attract the *north* needle in your compass, the pole that lies just under the surface at the Earth's "north magnetic pole" is really a south pole, not a north pole. But unlike labeling a bar magnet, the Earth's actual South Pole is always called its North Pole — for which we have compasses to thank.

The distance from the geographic North Pole to the Earth's magnetic north pole is fairly large — the magnetic north pole currently lies near Ellesmere Island in Northern Canada. Actually, the position of the Earth's magnetic north pole wanders over the years. The magnetic pole's yearly wanderings are appreciable: It's currently moving at a rate of more than 40 kilometers per year! The Earth's magnetic field is maintained by the swirling movements of the molten iron deep inside the Earth, in the planet's liquid outer core, allowing the magnetic pole to wander.

So how far away is the magnetic pole from the geographic pole? That's measured by the *angle of declination*. That angle varies, depending on your position on the Earth, but it can be sizeable. For example, in New York City, the angle of declination is about 12°. You can find out much more on the Earth's magnetic field, with current data and many interesting links, on the Web sites of the Geological Survey of Canada (`gsc.nrcan.gc.ca/geomag`) and the United States Geological Survey (`geomag.usgs.gov`).

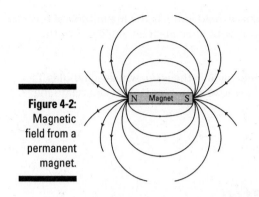

Figure 4-2:
Magnetic
field from a
permanent
magnet.

Note that the magnetic field from a magnet like the one in Figure 4-2 isn't
very constant or uniform — just like the electric field from two point charges
wouldn't be uniform.

If you want a uniform magnetic field, you usually select a location between
the two poles of a strong magnet, as Figure 4-3 shows. You can also create
a uniform magnetic field using coils of current, as I explain later in "Adding
loops together: Making uniform fields with solenoids."

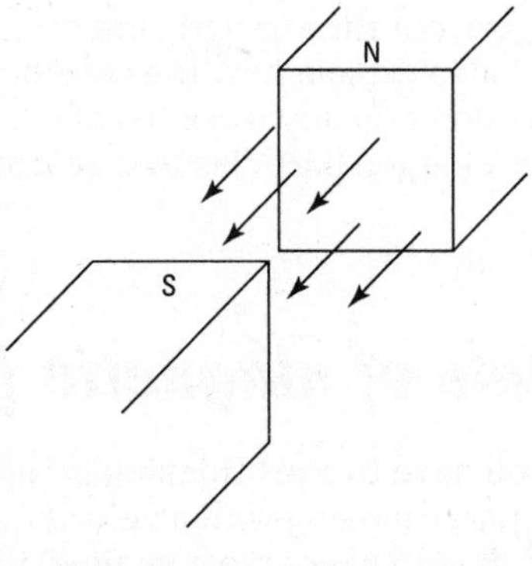

Figure 4-3:
A uniform
magnetic
field
between
two poles.

Defining magnetic field

Magnetism and electricity are so interconnected that *magnetic field* is defined
in terms of the strength of the force it exerts on a positive test charge. The
symbol *M* was already taken (it stands for the magnetization of a material),
so magnetic field ended up with the symbol *B*. Here's the formal definition of
magnetic field, from a physics point of view:

$$B = \frac{F}{qv\sin\theta}$$

Here, B is the magnitude of the magnetic field and F is the magnitude of the force on the charge q, which is moving with speed v at an angle of θ to the direction of the magnetic field.

In the MKS system, the unit of magnetic field is the *tesla,* whose symbol is T. In the CGS system, you use the *gauss,* whose symbol is G. You can relate the two like this:

$$1.0 \text{ G} = 1.0 \times 10^{-4} \text{ T}$$

Moving Along: Magnetic Forces on Charges

Electric currents and magnetic fields are linked very closely. Not only do electric currents give rise to magnetic fields, but magnetic fields also exert forces on the electric charges moving in currents.

Note that a charge has to be moving in order for a magnetic field to exert a force on it: No motion, no force on the charge.

In this section, I show you how to figure out the strength and direction of the magnetic force on a moving charge. I also explain how the direction of that force can ensure that magnetic fields don't do any work. To finish, you see why the direction of the force causes charged particles to travel in circles in a magnetic field.

Finding the magnitude of magnetic force

To get numerical with magnetism, you have to start thinking in terms of vectors. Suppose you have an electric charge moving with a velocity v. That charge is subject to a magnetic field, B. And of course, you need the F vector for the resulting force.

How can you determine the actual force, in newtons, on a charged particle moving though a magnetic field? That force is proportional to both the magnitude of the charge and the magnitude of the magnetic field. It's also proportional to the component of the charge's velocity that's *perpendicular to the magnetic field*. In other words, if the charge is moving along the direction of the magnetic field, parallel to it, there will be *no force* on that charge. If the charge is moving at right angles to the magnetic field, the force is at its highest.

Putting all this together gives you the equation for the magnitude of the force on a moving charge, where θ is the angle (between 0° and 180°) between the *v* and *B* vectors:

$F = qvB \sin \theta$

For example, suppose you're carrying around a 1.0-coulomb charge, and you experience a force from the Earth's magnetic field. The Earth's magnetic field on the surface is about 0.6 gausses, or 6.0×10^{-5} teslas. The faster you move with your charge, the more force you'll feel, so suppose you take it for a spin in a race car. Head off down the track straight at about 224 miles per hour, or 100 meters per second. What force do you feel on your charge at this speed in the direction perpendicular to the field? You know that the magnitude of the force is given by

$F = qvB \sin \theta$

So plug in the numbers. Here's what that looks like when you do:

$F = qvB \sin \theta$

$= (1.0 \text{ C})(100 \text{ m/s})(6.0 \times 10^{-5} \text{ T}) \sin 90°$

$= 6.0 \times 10^{-3} \text{ N}$

The force on the charge is 6.0×10^{-3} newtons, which is less than the weight of a paperclip.

Finding direction with the right-hand rule

Say you have a charge, *q,* traveling along with velocity *v,* minding its own business. If that charge travels in a magnetic field, *B,* there's going to be a force on the charge. You can see the direction of the magnetic force on the moving charge in Figure 4-4.

A right-hand rule operates when you're finding the force on a charge, and there are two versions of it — use whichever one you find easier:

✔ If you place the fingers of your open right hand along the magnetic field, the vector **B** in the figure, and your right thumb in the direction of the charge's velocity, **v,** then the force on a positive charge extends out of your palm (see Figure 4-4a). For a negative charge, reverse the direction of the force.

✔ Place the fingers of your right hand in the direction of velocity of the charge, **v,** and then wrap those fingers by closing your hand through the smallest possible angle (less than 180°) until your fingers are along the direction of the magnetic field, **B.** Your right thumb points in the direction of the force (see Figure 4-4b). For a negative charge, reverse the direction of the force.

Give the two methods a try to make sure you get the direction of the force correct.

Figure 4-4:
The force on a charge on a magnetic field and the associated right-hand rules.

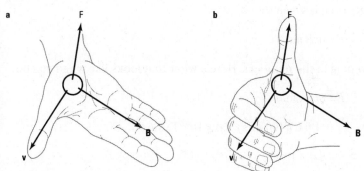

A lazy direction: Seeing how magnetic fields avoid work

Magnetic fields are lazy: They do no work on charged particles that travel through them — at least, not by the physics definition of work. So a charged particle in a magnetic field doesn't gain or lose kinetic or potential energy.

When you have an electric field, the situation is very different. There, the electric field pushes a charge *along or against* the direction of travel. And that's the physics definition of *work:*

$$W = Fs \cos \theta$$

where *F* is the force applied, *s* is the distance over which it's applied, and θ is the angle between the force and the direction of travel. In fact, that's where the whole idea of electric potential, *voltage,* comes from — the amount of work done on a charge divided by the size of the charge:

$$V = \frac{W}{q}$$

For the work done by a magnetic field, the trouble is the definition of work: $W = Fs \cos \theta$. The issue here is that in a magnetic field, the force and the direction of travel are always perpendicular to each other — that is, $\theta = 90°$ (see the preceding section). And $\cos 90° = 0$, so the work done by a magnetic field on a moving charge, $W = Fs \cos \theta$, is automatically zero.

That's it — the work done by a magnetic field on a moving charge is zero. That's why there's no such thing as magnetic potential (would that be *magnetic volts?*) to correspond to electric potential.

That's all due to the physics definition of work — work changes the kinetic or potential energy of a system (or the energy is lost to heat), and nothing of the kind happens with magnetic fields. However, the *direction* of the charged particle does change. That's what changes — not the particle's speed but its direction.

Going orbital: Following charged particles in magnetic fields

The direction and magnitude of the force in a magnetic field affects the path that an electric charge takes. The direction of the force causes the charge to move in circles, and the force's magnitude affects how big of a radius that circle has. In this section, I discuss the orbital motion of charges in magnetic fields.

Getting the curve

If you have an electric field (see Chapter 3), you know which way electric charges will move in such a field — along the electric field lines. For example, if you have a parallel plate capacitor, electrons will travel between the plates along the electric field lines, toward the positive plate. Protons will do the same, except they'll move toward the negative plate.

The situation changes when you have a magnetic field, not an electric field. Now the force is perpendicular to the direction of travel, which can take a little getting used to. To better show the path of the charge, physicists often draw the magnetic field as though you were looking at it straight on. How can you tell which way the magnetic field is going? Here's the physics way of showing direction:

- ✔ **Away from you:** If you see a bunch of *X*'s, the magnetic field goes into the page. Those *X*'s are intended to be the end of vector arrows, seen tail-on (imagine looking down the end of a real arrow, tail toward you).

- ✔ **Toward you:** Dots with circles around them are supposed to represent arrows coming at you, so in that case, the field is coming toward you.

Take a look at Figure 4-5, which shows the path a positive charge moving in a magnetic field will take. The positive charge travels along a straight line, undeflected, until it enters the magnetic field that goes into the page (represented by the *X*'s). Then a force appears on the charge at right angles, bending its path, as you can see in the figure.

Note: This is a good place to test your understanding of the right-hand rule of magnetic force (see the earlier section "Finding direction with the right-hand rule"). Apply it to the velocity and magnetic field you see in Figure 4-5 — do you agree with the direction of the resulting force?

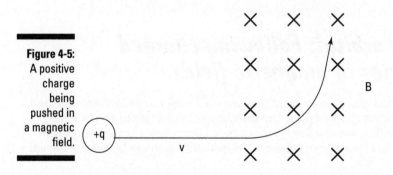

Figure 4-5:
A positive charge being pushed in a magnetic field.

Going in circles

Here's an interesting point: Which way do you get pushed if you're a charged particle moving in a magnetic field? The magnetic field is always perpendicular to the direction of travel (as Figure 4-4 shows earlier in this chapter). And no matter which way the charged particle turns, the force on it is perpendicular to its motion.

That's the hallmark of circular motion: The force is always perpendicular to the direction of travel. Therefore, charged particles moving in magnetic fields travel in circles.

See Figure 4-6 to get the full picture. There, a positive charge is moving to the left in a magnetic field. The dots with circles around them tell you that magnetic field B is coming straight at you, out of the page. Using the right-hand rule, you can tell which way the resulting force goes — upward when the positive charge is at the location in Figure 4-6.

What happens? The charge responds to that upward force by moving upward. And because the force due to the magnetic field is always perpendicular to the direction of travel, the force changes direction, too.

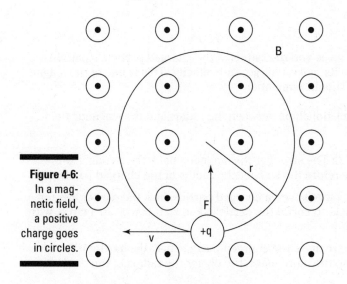

Figure 4-6:
In a mag-netic field, a positive charge goes in circles.

Finding the radius of orbit

Suppose you want to know the radius of the orbit of the charged particle moving in a magnetic field. Because the force is always perpendicular to the direction of travel, you end up with circular motion. And from Physics I, you have the following equation for the force needed to keep an object in circular motion:

$$F = \frac{mv^2}{r}$$

Here's the magnitude of the force on a charged particle moving in a magnetic field:

$$F = qvB \sin \theta$$

Because v is perpendicular to B in this case, θ equals 90°; therefore, $\sin \theta$ equals 1, which means you get this:

$$F = qvB$$

So set the two force equations — for circular motion and for the charged particle in the magnetic field — equal to each other:

$$qvB = \frac{mv^2}{r}$$

Rearranging this equation gives you this new version, solved for the radius:

$$r = \frac{mv}{qB}$$

That's great — that gives you the radius of the charged particle's path in a magnetic field, given its mass, charge, and velocity. This is one of the magnetism equations you should remember.

Note the following relationships between the radius and the magnetic field, mass, and velocity:

- **Magnetic field B:** The stronger the magnetic field, the stronger the force — and therefore the smaller the radius of the charged particle.

- **Velocity v:** The more speed a charged particle has, the harder it is for the magnetic field to corral the particle, and so it travels in a circle with a bigger radius.

- **Mass m:** The more mass the charged particle has, the harder it'll be to bend its path, so the more mass, the bigger the radius of the circle it travels in.

Notice how the equation reflects all these ideas: The magnetic field B is in the denominator of the fraction, so increasing B would give you a smaller answer for r; m and v are on top, so increasing either one of those would give you a larger r.

How about seeing this in action? Try some numbers. Say, for example, that you have a bunch of electrons going at 1.0×10^6 meters per second. You don't want to disturb the neighbors, so you decide to build a magnetic containment vessel to contain the electrons, sending them around in a circular orbit. Checking your bank account, you see you have only enough money to create a magnetic confinement vessel of $r = 1.0$ centimeters (even that may make your landlady suspicious, but she's learned that physicists sometimes need unusual equipment).

So what magnetic field do you need to limit your electrons to an orbit where $r = 1.0$ centimeters? You know that

$$r = \frac{mv}{qB}$$

An electron has a mass of 9.11×10^{-31} kilograms and a charge of 1.6×10^{-19} coulombs. Plugging in the numbers for electrons moving at 1.0×10^6 meters per second, you get

$$0.010 \text{ m} = \frac{\left(9.11 \times 10^{-31} \text{kg}\right)\left(1.0 \times 10^6 \text{m/s}\right)}{\left(1.6 \times 10^{-19} \text{C}\right)B}$$

Rearranging this equation and solving for B gives you the answer:

$$B = \frac{\left(9.11 \times 10^{-31} \text{kg}\right)\left(1.0 \times 10^6 \text{m/s}\right)}{\left(1.6 \times 10^{-19} \text{C}\right)\left(0.010 \text{m}\right)} \approx 5.7 \times 10^{-4} \text{T}$$

That's a very modest magnetic field — it's not much more than the Earth's magnetic field. It really doesn't take much to push an electron around. (That's a relief, because if you'd needed a magnetic field of several teslas, the landlady's silverware, which is silver-plated steel, might've ended up stuck to her ceiling.)

The equation

$$r = \frac{mv}{qB}$$

doesn't apply if the charged particle is traveling near the speed of light, $v \approx 3.0 \times 10^8$ meters per second, because relativistic effects take over, which affect the mass and orbital radius of the charged particle. I discuss special relativity in Chapter 12.

Selecting your atoms with a mass spectroscope

Mass spectroscopes, which are machines that determine which chemical elements go into a sample you're analyzing, rely on orbits in magnetic fields. A mass spectroscope heats the sample you want to analyze, ionizing some of the atoms. The singly ionized atoms get a net charge, *e* (the same magnitude as an electron's charge), and those atoms are accelerated through an electric potential, *V,* which gives them kinetic energy. The ionized atoms then enter a magnetic field, *B,* and they turn with radius *r.* You can pick them up with a detector, and positioning the detector tells you the radius for the ionized atoms — and from knowing the radius, electric potential, and magnetic field, you can determine the mass of the atoms and therefore identify them.

So given *r, e,* and *B,* you solve for *m,* the mass. The accelerating electric potential, *V,* gives

(continued)

(continued)

each ionized atom a kinetic energy, and the energy added by the electric potential must equal the kinetic energy added to each ion like this:

$$eV = \frac{1}{2}mv^2$$

Solving this equation for velocity, v, gives you

$$v = \left(\frac{2eV}{m}\right)^{\frac{1}{2}}$$

The radius of curvature of the ion in the magnetic field is

$$r = \frac{mv}{eB}$$

And solving for m in this equation gives you

$$m = \frac{reB}{v}$$

Because you just found v to be $\left(\frac{2eV}{m}\right)^{\frac{1}{2}}$, you can substitute for v:

$$m = \frac{reB}{\left(\frac{2eV}{m}\right)^{\frac{1}{2}}}$$

If you square both sides of this equation, you have

$$m^2 = \frac{r^2e^2B^2}{\left(\frac{2eV}{m}\right)}$$

Moving things around a little algebraically gives you the mass of the ionized atom:

$$m = \frac{er^2}{2V}B^2$$

And there you have it — the next time you come across a mass spectrometer, you'll know just how to find the masses of the atoms in any sample you put into it.

Down to the Wire: Magnetic Forces on Electrical Currents

You may be one of those rare physicists who doesn't have a bunch of electrons hurtling around at home at 1.0×10^6 meters per second. You may think that the preceding discussion, about charged particles in magnetic fields, doesn't really apply to you. However, you surely have electric cables around the house — and what are electric cables but wires through which charges move? In this section, you look at the forces that magnetic fields exert on charges moving in electric wires.

From speed to current: Getting current in the magnetic-force formula

To find the magnetic force on an electric wire in a magnetic field, you can start with the formula for individual charges. Take a look at the equation for the force on a moving electric charge in a magnetic field:

$$F = qvB \sin \theta$$

You want to translate this equation so that instead of using the speed of charged particles, v, it uses electrical current, I. How do you get electrical current out of this? *Current* is charge, q, divided by the amount of time, t, that a charge takes to pass a particular point:

$$I = \frac{q}{t}$$

Divide the equation for force by time and multiply it by time, which doesn't actually change the equation. Here's what you get:

$$F = \frac{q}{t}(vt)B\sin\theta$$

Note that you now have q/t, or current, here. So here's the force equation in terms of current:

$$F = I(vt)B\sin\theta$$

Okay, so what's $I(vt)B\sin\theta$? Something of a mixed bag here — current and speed together. But if you think about it, the term vt is just the speed of the charged particles making up the current multiplied by the measured time — and speed times time equals a distance. So replace vt with L, the distance the charged particles go in time t.

So here's the force on a wire of length L carrying current I in a magnetic field of strength B, where the L is at angle θ with respect to B:

$$F = ILB\sin\theta$$

Cool. How about an example? Look at Figure 4-7, which shows an electric current I in a magnetic field B. In physics, *current* goes in the direction a stream of positive charges would take (that convention was defined before scientists knew that it was negative charges — electrons — that really flowed to make current go). That means you can apply the right-hand rule to the situation you see in Figure 4-7 — just treat the direction of I as the direction a positive charge is traveling in. (For the right-hand rule, see the earlier section "Finding direction with the right-hand rule.")

All right, what if the current, I, equals 1.0 amp, and the magnetic field, B, is equal to 5 teslas? The magnetic force on a wire carrying this current increases in proportion to its length. For every meter of wire that you have, what would be the resulting force? You start with the formula for force:

$$F = ILB\sin\theta$$

and then divide by the length, L, to find the force per meter:

$$\text{Force/meter} = IB\sin\theta$$

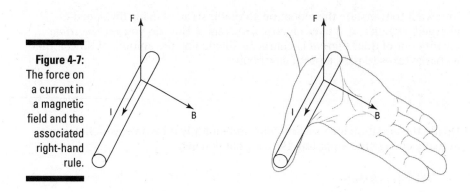

Figure 4-7:
The force on
a current in
a magnetic
field and the
associated
right-hand
rule.

In Figure 4-7, θ is 90°, and because sin 90° equals 1, you get this case:

Force/meter = IB

$= (1.0 \text{ A})(5.0 \text{ T}) = 5.0 \text{ N/m}$

So you get a result of 5.0 newtons per meter, which works out to be about a third of a pound per foot — something to keep in mind if you have electric cables running through a 5.0-tesla magnetic field (which, admittedly, is pretty rare).

Torque: Giving current a twist in electric motors

Scientists saw that magnetic fields exerted forces on electric wires and came up with electric motors. From there came electric washers and dryers, windshield wipers on cars, elevators, automatic doors at grocery stores, refrigerators, and much more (not in that order, of course). As you can see, life without electric motors would be inconvenient. This section helps you see what makes electric motors work, electricity and magnetism-wise — at least in basic terms.

Big-time currents

In the big physics labs, where cables can hold huge current (direct current, not alternating current), you can see something curious: When a cable is made up of individual strands of wire, those wires create magnetic field, and that magnetic field acts on the other wires in the cable. The net result is that the cable contracts before your very eyes, getting thinner as the magnetic fields act on the currents.

Seeing how motors work

Figure 4-8 shows an electric motor, stripped down to its basic components. Two permanent magnets of opposite polarity are on either side of the motor. This generates a uniform magnetic field in the space between the poles, from the north pole to the south pole. In this magnetic field, you place a loop of wire, which is free to rotate about the axis in the figure. A battery is connected to the loop, so a current is flowing through the wire in the direction shown by the arrows labeled *I*.

The wire loop is connected to the battery by a strange connection called a *commutator*. This clever little device is a vital part of the motor because it ensures that the current always flows in the direction shown in the diagram, even when the loop has rotated half a turn. It always connects the side of the loop that's closest to the north pole of the magnet to the positive terminal of the battery and vice versa, while leaving the loop free to turn.

Figure 4-8: Forces, current, and magnetic field in an electric motor.

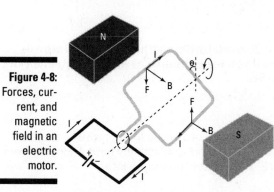

Because the loop is carrying current, the loop experiences a force in the magnetic field. I've shaped the loop as a rectangle to make the calculation of the force it experiences a little easier.

Two sides of the loop are parallel to the axis of rotation, and two are perpendicular to it. The perpendicular wires don't play a part here because the force they experience is directed along the axis of rotation, so they don't produce any torque. Also, they're equal and opposite in size, so you don't get a net force from them.

Most interesting are the two parts of the loop that run parallel to the axis of rotation, which are always at 90° to the magnetic field. The left side of the wire loop is forced down, and the right side is forced up (you can use the right-hand rule to confirm that the directions of the forces in Figure 4-8 are correct). This results in a turning force — that is, a *torque* — that rotates the loop of wire. If you connect the loop to an axle, then the loop will force the axle to turn — and you can use this turning force for all sorts of things.

Figuring out the turning force

So how much turning force does an electric motor give you? *Torque,* as you may recall from Physics I, is a twisting force, with the symbol τ. Here's the formula for it:

$$\tau = Fr \sin \theta$$

where F is the applied force, r is the distance the force acts from the turning point, or *pivot,* and θ is the angle between F and r.

In an electric motor, a loop of current is embedded in a magnetic field, B, and that field creates forces, F, on each wire running parallel to the axis of rotation (as you see in Figure 4-8). The torque on each wire is the force $(F = ILB)$, multiplied by the distance, d, the force acts from the pivot multiplied by the sine of the angle. Because there are two torques, corresponding to the two sides of the loop, the total torque, τ, is equal to the following:

$$\tau = ILB\left(\frac{1}{2}d \sin\theta + \frac{1}{2}d \sin\theta\right) = ILBd \sin\theta$$

The product dL is the height multiplied by the width of the loop of wire — that is, the *area* of the loop. So if you write the area as A, the equation for the torque on a loop of wire becomes

$$\tau = IAB \sin \theta$$

In fact, electric motors aren't really made of a single loop of wire — they're made of coils of wire. So instead of one loop, you actually have N loops, where N is the number of coils of wire. That makes the torque into

$$\tau = NIAB \sin \theta$$

That's the total torque on a coil of N loops of wire, each carrying current I, of cross-sectional area A, in a magnetic field B, at angle θ as shown in Figure 4-8.

In physics class, you're usually asked what the maximum torque would be for such-and-such a coil in such-and-such a magnetic field. If you come across a situation like that and need to find the maximum torque, that occurs when the coil is at right angles to the magnetic field: $\theta = 90°$, so $\sin \theta = 1$, or

$$\tau = NIAB$$

Try some numbers here. If you have a coil with 200 turns, a current of 3.0 amps, an area of 1.0 square meters, and a magnetic field of 10.0 teslas, what's the maximum possible torque? Just plug this into the equation:

$$\tau = NIAB = (200)(3.0 \text{ A})(1.0 \text{ m}^2)(10.0 \text{ T}) = 6.0 \times 10^3 \text{ N-m}$$

So you have a maximum torque of 6,000 newton-meters, which is very large — a car typically generates only about 150 newton-meters.

Going to the Source: Getting Magnetic Field from Electric Current

The earlier sections in this chapter concentrate on how magnetic fields exert forces on moving charges, or currents, without worrying too much about where the magnetic field came from in the first place. In this section, you discover the source of that magnetic field. Here, you see the relationship between electricity and magnetism become complete.

Simply put, just as electric charges are the source of electric fields, which exert forces on other electric charges, electric currents are the source of the magnetic fields, which exert forces on other electric currents.

In Chapter 3, I take a couple of simple arrangements of charge (the point charge and the parallel plate capacitor) and examine the resulting electric fields. Now, in this section, I take a few simple arrangements of current (a straight wire, a loop, and a tube of current called a *solenoid*) and examine the resulting magnetic fields. Here, you also see how you can use this idea to make *electromagnets,* magnets that you can switch on and off with a switch.

Producing a magnetic field with a straight wire

To understand how electric current produces a magnetic field, first take a look at the magnetic field from a single wire, as Figure 4-9 shows. Why start here? When you know what the magnetic field is from a single wire of current, you're home free in many problems. You can often break down more-complex distributions of current into many single wires — and then add the magnetic fields from the wires as vectors to get the overall result.

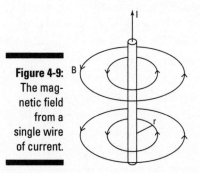

Figure 4-9: The magnetic field from a single wire of current.

Assembling the formula for magnetic field from a single wire

When you make physical measurements of the magnetic field from a single wire, you find that the magnetic field, B, diminishes as 1 over the distance, r. Therefore, you get this relation (where $\propto$ means *proportional to*):

$$B \propto \frac{1}{r}$$

What else can the magnetic field depend on? Well, how about the current itself, I? Surely if you double the current, you get twice the magnetic field, right? Yep, that's the way it works, as borne out by measurement, so now you have the following:

$$B \propto \frac{I}{r}$$

That's all you need.

The constant of proportionality, for historical reasons, is written as $\mu_o/(2\pi)$, which means you finally get this result for the magnetic field from a single wire:

$$B = \frac{\mu_o I}{2\pi r}$$

Note that the constant $\mu_o = 4\pi \times 10^{-7}$ T-m/A.

A right-hand rule: Finding field direction from a wire

Magnetic field, B, is a vector. If you have a magnetic field from the current in a single wire, which way does the B field go? There's another right-hand rule for just this occasion. If you put the thumb of your right hand in the direction of the current, the fingers of that hand will wrap around in the direction of the magnetic field. At any one point, the direction your fingers point is the direction of the magnetic field, as Figure 4-10 shows.

Figure 4-10:
A right-hand rule shows the direction of current in a wire and the resulting magnetic field.

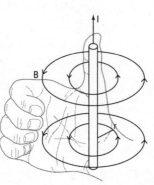

Give this a try: Suppose you have two parallel wires. Verify that the force between two wires is toward each other if the current in both is in the same direction and away from each other if the current in each wire is in opposite directions.

Calculating magnetic field from straight wires

How about some numbers? Say you have a current of some 10 amps and you want to measure the magnetic field 2.0 centimeters from the center of the wire. What is the strength of the B field you'll get? Here's your formula:

$$B = \frac{\mu_o I}{2\pi r}$$

Plugging in the numbers gives you:

$$B = \frac{\mu_o I}{2\pi r} = \frac{\left(4\pi \times 10^{-7}\,\text{T•m/A}\right)(10\,\text{A})}{2\pi\,(0.020\,\text{m})} \approx 1.0 \times 10^{-4}\,\text{T} = 1.0\,\text{G}$$

So you get 1.0 gausses, a little more than the Earth's magnetic field, which is about 0.6 gausses.

That was a quick example — how about one that's a little tougher? Say you have two wires, parallel to each other, with current I in each going the same way. The wires are a distance r apart. What's the force on Wire 1 from the magnetic field coming from Wire 2?

You know that the force on Wire 1, which is carrying current I in magnetic field B, is the following (to see where this formula comes from, check out the earlier section "From speed to current: Getting current in the magnetic-force formula"):

$$F = ILB$$

All right, but what's B? That's the magnetic field from Wire 2, measured at the position of Wire 1. Because the wires are r distance apart and Wire 2 is carrying a current I, its magnetic field is this at the location of Wire 1:

$$B = \frac{\mu_o I}{2\pi r}$$

Substituting this expression for B into the $F = ILB$ equation, you get this result:

$$F = \frac{\mu_o I^2 L}{2\pi r}$$

How about getting the force per unit length? That's F/L, which is

$$\frac{F}{L} = \frac{\mu_o I^2}{2\pi r}$$

Now try some numbers. Say you have two parallel wires with current I going in the same direction — current, I, is 10 amps, and the distance between the wires, r, is 2.0 centimeters. Putting in those numbers, you get

$$\frac{F}{L} = \frac{\mu_o I^2}{2\pi r} = \frac{\left(4\pi \times 10^{-7}\,\text{T·m/A}\right)\left(10\,\text{A}^2\right)}{2\pi\left(0.020\,\text{m}\right)} \approx 1.0 \times 10^{-3}\,\text{N/m}$$

So the force on Wire 1 from Wire 2 is 1.0×10^{-3} newtons per meter. Note that the force on Wire 2 from Wire 1 is the same magnitude.

Getting centered: Finding magnetic field from current loops

Suppose you have a loop of current, such as you see in Figure 4-11. The magnetic field from a single loop of wire (even if it has many turns) is not constant over the various points in space.

That variation in the magnetic field is a bit of a problem, because the actual equation for the magnetic field from a loop of current is very complicated. So physicists do what they always like to do — they simplify. Here, simplifying takes the form of measuring the magnetic field at the very center of the loop. (In the next section, you see that putting multiple loops together to form a tube of current also smooths out the magnetic field.)

Here, start by noting that the magnetic field at the center of a loop of current is equal to the following:

$$B = N\frac{\mu_o I}{2R}$$

where N is the number of turns in the loop, I is the current in the loop, and R is the radius of the loop.

TIP

What's the direction of the B field at the center of the loop of wire? You guessed it — there's a right-hand rule for that. To apply this rule, just wrap the fingers of your right hand around the loop in the direction the current is going — your right thumb points in the direction that the B field points in the center of the loop.

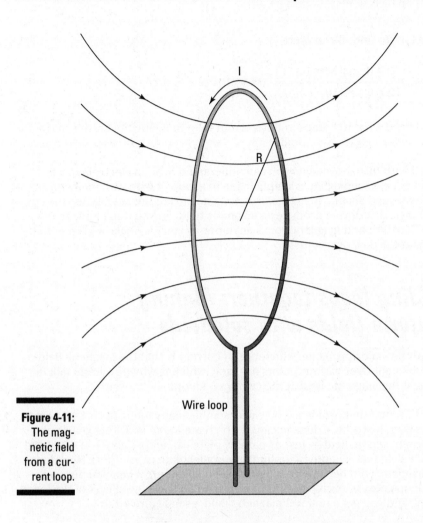

I

R

Wire loop

Figure 4-11:
The mag-
netic field
from a cur-
rent loop.

Try some numbers. Say that you have a loop of 200 turns of wire and a radius
of 10 centimeters. What current would you need to get the equivalent of the
Earth's magnetic field, 0.6 gausses, in the center?

Plug in the numbers, making sure you first convert from gausses to teslas
($1.0 \text{ G} = 1.0 \times 10^{-4}$ T) and from centimeters to meters. Here's what you get:

$$B = N \frac{\mu_o I}{2R}$$

$$6.0 \times 10^{-5} \text{T} = \frac{(200)\left(4\pi \times 10^{-7} \text{T·m/A}\right)I}{2(0.10 \text{ m})}$$

Solve for I to find the answer:

$$I = \frac{2(0.10 \text{ m})(6.0 \times 10^{-5} \text{ T})}{(200)(4\pi \times 10^{-7} \text{ T} \cdot \text{m/A})} \approx 4.8 \times 10^{-2} \text{ A}$$

You'd need 4.8×10^{-2} amps in this current loop to mimic the Earth's magnetic field.

Armed with this knowledge, you can understand how an electromagnet works. An electromagnet is simply made of a loop of wire with many turns, usually wound around a piece of iron to concentrate the field. When the current flows, this device produces a magnetic field. So you don't have to dig in the Earth to find magnetic rocks anymore — you can make a magnet that works at the flick of a switch.

Adding loops together: Making uniform fields with solenoids

One of the major problems with loops of current is that the magnetic field isn't constant over various points in space, which is why physicists talk in terms of the magnetic field at the center of a loop.

To get around that problem, you can assemble many loops of current next to each other, just a little distance apart, to create a *solenoid*. This gives you a uniform magnetic field — just as parallel plate capacitors give you a uniform electric field (see Chapter 3 for info on parallel plate capacitors). How does the magnetic field become constant inside a solenoid? When you put multiple loops next to each other, as in Figure 4-12a, the edge effects of the loops cancel, and you get a uniform magnetic field, as in Figure 4-12b.

What is the magnitude of a solenoid's magnetic field? If the length of the solenoid is large compared to its radius, you get this equation for the magnetic field:

$B = \mu_o n I$

where n is the number of wire loops in the solenoid per meter — that is, the number of turns per meter — and I is the current in each turn.

How about the direction of the magnetic field? That's easy enough: You can use the right-hand rule for current loops (see the preceding section) to figure out which direction the magnetic field goes in for a solenoid. Just take a look at Figure 4-12 to confirm you're getting it right.

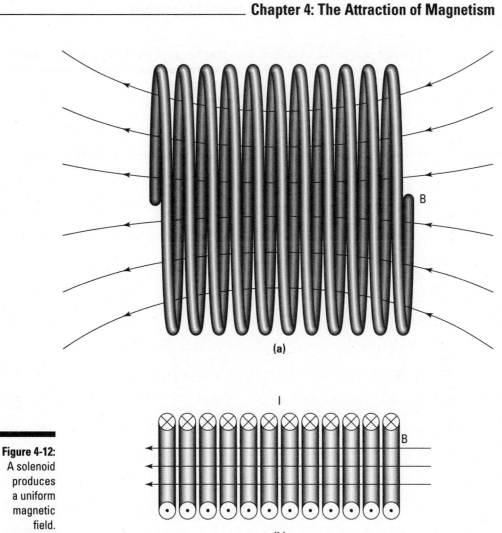

(a)

(b)

Here's an example. Say that you're conducting some crucial lab experiments and need a 1.395-tesla magnetic field. How much current would you need to run through a solenoid of some 3,000 loops, 1.00 centimeters in length, to get that magnetic field?

Start by solving for the current, I:

$$I = \frac{B}{\mu_o R}$$

Then just plug in the numbers; note that because you have 3,000 1-centimeter loops, you use 300,000 — or 3.0×10^5 loops per meter — as your value for n:

$$I = \frac{B}{\mu_o n} = \frac{1.395 \text{ T}}{\left(4\pi \times 10^{-7} \text{ T} \cdot \text{m/A}\right)\left(3.00 \times 10^5\right)} \approx 3.70 \text{ A}$$

In other words, you need about 3.70 amps, which isn't too much.

Chapter 5

Alternating Current and Voltage

In Physics I, you work with direct-current (DC) circuits, where the current is driven by a battery. Here, you take a look at alternating current (AC) in circuits. Things get more active, because you're dealing with *alternating voltages,* which means that the voltage in any wire changes from positive to negative and then back again regularly.

You may wonder why alternating current is such a big deal. Many types of circuits, including those you tune to receive signals from airborne waves, would be impossible without it. But alternating current got its start in a big way when people first started sending electricity through power lines. Direct current, which doesn't alternate, could travel only a short distance before the resistance of the wires overcame the current. Alternating current, however, can travel much farther with no problem (it actually helps regenerate itself through alternating magnetic and electric fields). That's why power lines always carry alternating current.

You see three types of circuit elements in this chapter: resistors, capacitors, and inductors. They all react differently to alternating current. It's going to be quite a ride, but I offer a guided tour of the whole shebang, so sit back and relax.

AC Circuits and Resistors: Opposing the Flow

Resistors are the easiest components to handle when dealing with AC circuits, perhaps because resistors don't care a bit if the current through them is alternating or not — they react in exactly the same way to alternating and direct voltages.

A *resistor* is a circuit element that literally resists current to some degree. Here's how it works: As I explain in Chapter 3, when there's a potential difference across a metal, for instance, the electric field induces a current by making the negatively charged electrons flow. As the electrons flow and barge their way past the atoms, they jostle the atoms, so the electrons encounter a resistance to their progress. Here are a couple of important points:

✔ The larger the potential difference you put across the metal, the stronger the electric field and the greater the current.

✔ The greater the resistance of the resistor, the less current you get for a given potential difference across it.

In an *ideal* resistor, the current is proportional to the potential difference. The size of the potential difference you need to make 1 unit of current flow is called the *resistance*.

In this section, you see how current, voltage, and resistance relate through Ohm's law for AC circuits. I also show you how voltage and current relate graphically when you have a resistor in an AC circuit.

Finding Ohm's law for alternating voltage

The current through a resistor is related to the voltage across the resistor by Ohm's law:

$$I = \frac{V}{R}$$

I is current, *V* is voltage, and *R* is resistance measured in ohms (Ω). So if you know the voltage across a resistor, you can find the current through it. Simple. Now take this picture from direct current to alternating current. To do that, upgrade from batteries — that is, sources of constant voltage — to alternating voltage sources.

The voltage from an alternating voltage source is not constant — it usually varies like a sine wave. You can see what the voltage from an alternating voltage source looks like in Figure 5-1. The *peak voltage* — that is, the maximum voltage — is equal to V_o.

Here's the formula for the voltage from an alternating voltage source as a function of time:

$$V = V_o \sin(2\pi f t)$$

Here, V_o is the maximum voltage and *f* is the frequency of the alternating voltage source. Frequency is measured in hertz, whose symbol is Hz. *Frequency* is the number of complete cycles (from peak to peak along the sine wave, for example) that occur per second.

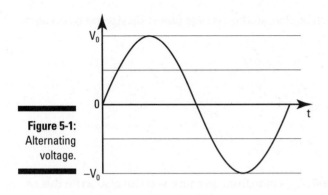

Figure 5-1:
Alternating
voltage.

How does this alternating voltage affect Ohm's law? Not too badly — Ohm's law just becomes

$$I = \frac{V_o}{R}\sin(2\pi ft)$$

You can rewrite Ohm's law in terms of the maximum current, I_o, like this (because $V_o = I_o/R$). So here's Ohm's law for an AC circuit:

$$I = \frac{V_o}{R}\sin(2\pi ft) = I_o\sin(2\pi ft)$$

Averaging out: Using root-mean-square current and voltage

When discussing AC circuits, you usually don't work in terms of maximum voltages and currents, V_o and I_o; instead, you speak in terms of the *root-mean-square* voltages and currents, V_{rms} and I_{rms}. What's the difference?

Root-mean-square is a way of treating circuits with alternating voltages much as you'd treat circuits with direct, nonalternating voltages. For example, here's what the power dissipated as heat in a circuit with nonalternating voltage looks like:

$$P = IV$$

And here's what the dissipated power looks like in a circuit with alternating voltage:

$$P = I_o V_o \sin^2(2\pi ft)$$

Not exactly the same, are they? So physicists talk in terms of the *average power* dissipated by a circuit with an alternating current source — that is, averaged over time. That's a way of looking at alternating-voltage circuits much as you'd look at battery-driven circuits. The time average of $\sin^2(2\pi ft)$

works out to be ½, which is nice, so the average power dissipated by an alternating voltage circuit is

$$P_{avg} = \frac{I_o V_o}{2}$$

You can also write this as

$$P_{avg} = \frac{I_o}{\sqrt{2}} \frac{V_o}{\sqrt{2}}$$

And that's where I_{rms} and V_{rms} come from, because you can also write this as

$$P_{avg} = I_{rms} V_{rms}$$

where $I_{rms} = \frac{I_o}{\sqrt{2}}$ and $V_{rms} = \frac{V_o}{\sqrt{2}}$.

So I_{rms} and V_{rms} are each the maximum current or voltage, divided by the square root of 2.

Staying in phase: Connecting resistors to alternating voltage sources

Say that you connect an alternating voltage source to a resistor, as Figure 5-2 shows. The circle around the ~ symbol represents an alternating voltage source, and the zigzag represents the resistor.

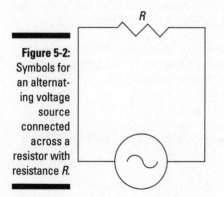

Figure 5-2:
Symbols for
an alternat-
ing voltage
source
connected
across a
resistor with
resistance R.

The voltage across the resistor is just the voltage supplied by the alternating voltage source, so the current through the resistor, at time *t*, is given by Ohm's law:

$$I = \frac{V_o}{R}\sin(2\pi ft) = I_o \sin(2\pi ft)$$

Note that if you square both sides of this current-voltage relationship and then take the average (remember that the average of $\sin^2(2\pi ft)$ works out to be ½), then you have a relation between the mean-squared voltage and current. If you take the square root, you get the following relation between the root-mean-square voltage and current across a resistor:

$$V_{rms} = I_{rms}\,R$$

This is the root-mean-square equivalent of Ohm's law in an AC circuit.

You can see a graph of the current and voltage across the resistor in Figure 5-3. Note that the current through the resistor and the voltage across the resistor rise and fall at the same time. That means that the current and the voltage in a resistor are *in phase*. (However, the current and voltages through and across capacitors and inductors do not mirror each other — that is, they're not in phase, as you see later in this chapter.)

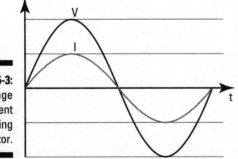

Figure 5-3:
Voltage
and current
alternating
in a resistor.

AC Circuits and Capacitors: Storing Charge in Electric Field

A *capacitor* is a device that stores charge when you apply a voltage across it. You may have already met the capacitor in Chapter 3, in the form of two parallel plates. The more charge you put on the plates, the greater the potential difference between them.

Generally, for any type of capacitor, the amount of charge stored for every unit of potential difference is called the *capacitance* (measured in *farads,* a unit named after Michael Faraday). The voltage across a capacitor (V) that has capacitance C is related to the amount of charge stored on it (Q) by the following equation:

$$V = \frac{Q}{C}$$

How does a capacitor react to alternating voltage? That's what you look at in this section.

Introducing capacitive reactance

Suppose you connect a capacitor to an alternating voltage source, as Figure 5-4 shows (the symbol for a capacitor is two upright bars, meant to represent the plates of a parallel plate capacitor).

Figure 5-4:
An alternating voltage source connected across a capacitor with capacitance C.

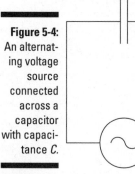

Here's how voltage relates to current when you have a capacitor and an alternating voltage source:

$$V_{\text{rms}} = I_{\text{rms}} X_{\text{c}}$$

where V_{rms} and I_{rms} are the *root-mean-square* voltage and current (the maximum voltage and current divided by the square root of 2 — see the earlier section "Averaging out: Using root-mean-square current and voltage" for details). Here, X_{c} is called the *capacitive reactance,* and it's equivalent to the resistance, R, in the root-mean-square voltage and current relation for the resistor (see the earlier section "Staying in phase: Connecting resistors to alternating voltage sources"). X_c is measured in ohms (Ω), just as R is, and it's equal to the following:

$$X_c = \frac{1}{2\pi f C}$$

where f is the frequency of the alternating voltage source and C is the capacitor's capacitance, measured in farads (F).

You can think of the *capacitive reactance* as the effective resistance the capacitor puts in the way of the alternating voltage source, much like R for resistors.

Note that the capacitive reactance depends on frequency, which is something that resistance doesn't do. When the frequency (f) is low, the capacitive reactance (X_c) is large, and when the frequency is high, the capacitive reactance is small. (The equation shows this relationship by putting f in the bottom of the fraction.)

Why is capacitive reactance high when the frequency is low and low when the frequency is high? Intuitively, you can think of it this way: When the frequency is high, the capacitor doesn't have much time between voltage reversals to accumulate new charge, so it doesn't change the voltage across it much. When the frequency is low, the capacitor has more time to accumulate charge during each cycle and so can change its voltage more.

How about seeing some numbers? Say you have a 1.50-µF capacitor (where µF is a microfarad, or 10^{-6} F), and you connect it across a voltage source whose V_{rms} = 25.0 volts. What is I_{rms} if the frequency of the voltage source is 100 hertz?

First, find the capacitive reactance:

$$X_c = \frac{1}{2\pi f C} = \frac{1}{2\pi(100 \text{ Hz})(1.50\times10^{-6} \text{ F})} \approx 1{,}060 \ \Omega$$

So the capacitive reactance is 1,060 ohms. Now put that to work finding I_{rms}. You know that $V_{rms} = I_{rms} X_c$, so

$$I_{rms} = \frac{V_{rms}}{X_c}$$

Plug in the numbers and solve:

$$I_{rms} = \frac{25.0 \text{ V}}{1{,}060 \ \Omega} \approx 2.36\times10^{-2} \text{ A}$$

And there you have the current — 2.36×10^{-2} amps. If the frequency were higher, the capacitive reactance would be lower, so the current would be higher (because the capacitive reactance is the effective resistance).

Getting out of phase: Current leads the voltage

When you have a capacitor in an AC circuit, the current and voltage sine waves are *out of phase* — that is, they're shifted in time with respect to each other: One reaches its peak before the other. You can see the applied voltage as a function of time in Figure 5-5 — as well as the actual current that flows in the circuit.

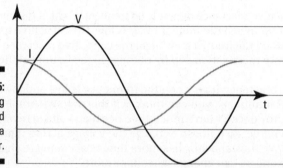

Figure 5-5:
Alternating voltage and current in a capacitor.

Notice that with a capacitor, the current *leads* the voltage — that is, the current reaches its peak before the voltage does. In fact, the current leads the voltage by exactly 90°, or π/2 radians — that is, one quarter cycle. So when you're graphing the current and voltage from a capacitor in an alternating voltage circuit, always remember that the current leads.

Why does current reach its peak before voltage does? The answer is simple if you think about how a capacitor works. Current piles charge onto a capacitor, so as long as the current is positive, the capacitor's voltage increases. When the current is decreasing in magnitude, it will still be positive for a while, so charge is still being added onto the capacitor; thus, the voltage keeps on increasing. Not until the current changes direction and goes negative does the charge start to come off the capacitor, causing the voltage to decrease. Therefore, the voltage reaches its peak after the current does.

In fact, you say that if the applied voltage is $V = V_0 \sin(2\pi ft)$, then current, which leads voltage by π/2 radians, looks like this — note that its argument (in parentheses) reaches a specific value before the voltage does because you're adding π/2 to 2πft:

$$I = I_0 \sin\left(2\pi ft + \frac{\pi}{2}\right)$$

Using a little trig, this becomes the following:

$$I = I_o \cos(2\pi ft)$$

So you can see that if the voltage goes as the sine, current goes as the cosine, so they're out of phase.

Preserving power

Here's something surprising: The average power dissipated by the capacitor is *zero*. Why? Well, the power used by an electrical component is $P = IV$. Here's what this looks like for a resistor, where the current and the voltage are in phase:

$$P = I_o V_o \sin^2(2\pi ft)$$

However, for a capacitor, the power looks like this, because the current and voltage are 90° out of phase:

$$P = I_o V_o \sin(2\pi ft) \cos(2\pi ft)$$

The time average of $\sin(2\pi ft) \cos(2\pi ft)$ is zero (because this product spends as much time positive as it does negative), so the average power used by a capacitor is zero:

$$P_{avg} = 0$$

This means that no power is lost to the environment as heat (as is the case with a resistor), and in fact, the capacitor spends as much time feeding power back into the circuit as it does getting power from the circuit: The capacitor feeds power back to the circuit when it's discharging and its voltage is going down, and the capacitor gains energy when it's being charged up and its voltage is increasing.

AC Circuits and Inductors: Storing Energy in Magnetic Field

Just as capacitors store energy in an electric field (that is, charges are separated by some distance, giving rise to an electric field), so inductors store energy — but this time, it's stored in a magnetic field. For example, a solenoid (see Chapter 4) is an inductor, because when you run current through it, a magnetic field appears — and doing that takes energy. In fact, the electrical symbol for an inductor is just that: a *solenoid,* or loops of current, as you can see in the circuit in Figure 5-6.

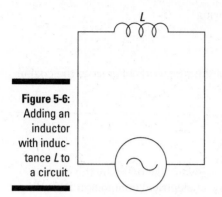

Figure 5-6:
Adding an
inductor
with induc-
tance *L* to
a circuit.

Inductors do the same kind of thing as capacitors: They shift the current relative to voltage from an alternating voltage source. However, instead of leading by $\pi/2$, the current lags by $\pi/2$.

For a capacitor (see the preceding section), the voltage is a function of the capacitance and the charge stored on one plate (the charge stored on one plate is equal in magnitude but opposite in sign to the charge stored on the other plate):

$$V = \frac{Q}{C}$$

A similar relationship exists for inductors, as you see in this section. Here, I show you how inductors produce a voltage based on the concept of Faraday's law. For that, I introduce the concept of magnetic flux, which you get when a magnetic field passes through a loop of wire. I also explain how inductive reactance, just like capacitive reactance, opposes an alternating current — only this time, voltage comes out ahead of current.

Faraday's law: Understanding how inductors work

Michael Faraday (the same physicist that *farads,* the units of capacitance, are named after), came up with *Faraday's law,* which says the following:

> When an inductor suffers a change in magnetic flux, it produces a voltage that tends to resist the change.

What does all that mean? This section explains the physics behind inductors, starting with the idea of magnetic flux.

Introducing flux: Magnetic field times area

When a magnetic field goes through a loop of wire, there's said to be a *magnetic flux* over the area of the loop. You can see how this works in Figure 5-7. There, a uniform magnetic field *(B)* is going through a wire loop with area *A*, which is orientated at an angle θ to the magnetic field.

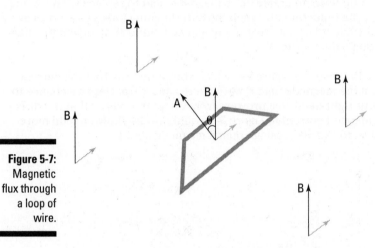

Figure 5-7: Magnetic flux through a loop of wire.

Here's where it gets strange: In physics, areas are often represented by vectors, and they point directly out of the flat surface whose area they represent. In other words, the vector *B* in Figure 5-7 should be familiar — that's just the magnetic field. But the area vector, *A,* is new — that's the vector that's perpendicular to the wire loop, and its magnitude is the same numerical value as the area of the loop.

Magnetic flux is the strength of the magnetic field multiplied by the component of the area vector parallel to the *B* field. In other words, magnetic flux is a magnetic field strength multiplied by an area. When the magnetic field is parallel to the area vector, the magnetic flux, whose symbol is Φ, is *BA*. On the other hand, when *B* is perpendicular to *A,* no field lines actually go through the wire loop, and the flux is zero. Putting all this together, here's what magnetic flux is in terms of *B, A,* and θ, the angle between them:

$$\Phi = BA \cos \theta$$

Inducing a voltage to keep the status quo

Faraday's law says that if the magnetic flux changes, it induces a voltage around the loop; that voltage creates a current in a way that opposes the change by creating its own magnetic field.

For instance, say that the magnetic field is decreasing in strength. The wire loop wants to keep things the way they are, so it resists change. The wire loop creates a voltage in itself that causes a current to flow — and that current creates a magnetic field.

The magnetic field is created in such a way as to preserve the status quo; thus, the current flows in a way that adds magnetic field to the applied magnetic field — that is, the applied magnetic field is decreasing, so the current around the loop flows to create more magnetic field to replace what's being decreased. (The inductor can't keep the current going forever — it dies away quickly, but while it lasts, it creates a magnetic field to supplement the magnetic field that's decreasing.)

You can see the result in Figure 5-8, which shows the way that the current would flow if the magnetic field *B* were decreasing. (*Tip:* Here's a chance to show off your right-hand rule prowess from Chapter 4. Verify that the direction of the induced current in Figure 5-8 would flow as shown to add more magnetic field to the decreasing applied magnetic field.)

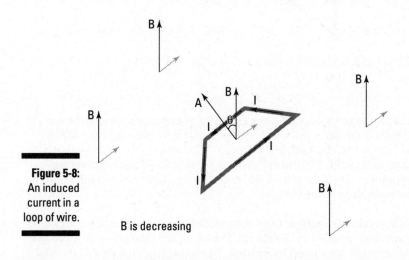

Figure 5-8:
An induced
current in a
loop of wire.

B is decreasing

Finding induced voltage using the change in magnetic flux

How is the voltage induced around the loop of wire related to the change in magnetic flux? The voltage looks like this:

$$V = -N\frac{\Delta\Phi}{\Delta t}$$

That is, the induced voltage is equal to the number of turns in the wire loop (*N*) multiplied by the change in flux ($\Delta\Phi$) divided by the time in which the change in flux takes place (Δt). The negative sign is there to remind you that the induced voltage acts to oppose the change in flux.

Try some numbers here. Say that you have a solenoid consisting of 40 turns of wire, each with an area of 1.5×10^{-3} square meters. A magnetic field of 0.050 teslas is perpendicular to each loop of wire (that is, $\theta = 0°$). A tenth of a second later, $t = 0.10$ s, the magnetic field has increased to 0.060 teslas. What is the induced voltage in the solenoid?

Start by finding the change in flux over a tenth of a second. The flux looks like this for each turn of wire:

$$\Phi = BA \cos \theta$$

Therefore, the original flux through each turn of wire is this, bearing in mind that $\theta = 0°$ and the original B field is B_o:

$$\Phi_o = B_o A$$

Putting in numbers gives you the following:

REMEMBER

$$\Phi_o = (0.050 \text{ T})(1.5 \times 10^{-3} \text{ m}^2) = 7.5 \times 10^{-5} \text{ Wb}$$

Wb stands for *weber*, the MKS unit of magnetic flux; it's equal to 1 T-m^2.

And the final magnetic flux is equal to this, where B_f is the final magnetic field:

$$\Phi_f = B_f A$$

Plugging in the numbers gives you the following:

$$\Phi_o = (0.060 \text{ T})(1.5 \times 10^{-3} \text{ m}^2) = 9.0 \times 10^{-5} \text{ Wb}$$

So the change in flux is

$$\Delta\Phi = \Phi_f - \Phi_o$$
$$= 9.0 \times 10^{-5} \text{ Wb} - 7.5 \times 10^{-5} \text{ Wb}$$
$$= 1.5 \times 10^{-5} \text{ Wb}$$

This change takes place in 0.10 seconds, and it takes place in all 40 turns of the solenoid, so the equation $V = -N\dfrac{\Delta\Phi}{\Delta t}$ becomes

$$V = -40\frac{1.5 \times 10^{-5} \text{ Wb}}{0.10 \text{ s}} = -6.0 \times 10^{-3} \text{ V}.$$

So there you have it — the voltage the solenoid creates to oppose the change in magnetic flux is 6.0 mV (millivolts). That's what the induced voltage starts off at — it decays exponentially in time.

Finding induced voltage using the change in current

The voltage induced by an inductor looks like this:

$$V = -N\frac{\Delta\Phi}{\Delta t}$$

However, if you have an *electrical inductor* — that is, a component in a circuit — you don't typically talk in terms of the change in flux inside that component. Instead, you talk about the change in current through the inductor, because that makes more sense in the context of circuits than speaking of magnetic flux.

How do you relate current through the solenoid and magnetic flux? Plugging in $\Phi = BA\cos\theta$ gives you the following:

$$V = -N\frac{\Delta(BA\ \cos\theta)}{\Delta t}$$

And for a solenoid, $B = \mu_0 nI$, where n is the number of wire loops in the solenoid per meter — that is, the number of turns per meter — μ_0 is $4\pi \times 10^{-7}$ T•m/A, and I is the current in each turn (see Chapter 4 for details). Also, because you have only one selenoid, with n turns per meter, then $N = 1$. So you can write the voltage as

$$V = -\frac{\Delta(\mu_0 nIA\ \cos\theta)}{\Delta t}$$

If the current is the only thing changing in an inductor that's part of an electric circuit, you get this:

$$V = -\frac{\mu_0 nA\ \cos\theta\ \Delta I}{\Delta t}$$

You wrap $\mu_0 nA\cos\theta$ up into one number — the *inductance* of the inductor, whose symbol is L, and whose units are *henries* (which my friend Henry thinks is a good idea). So you have this equation to tie the induced voltage to the change in current over time:

$$V = -L\frac{\Delta I}{\Delta t}$$

That's the result you're looking for — the inductance connects the change in current over time to the induced voltage. And so all inductors you see in circuits are labeled with their inductance in henries (H).

Introducing inductive reactance

For a resistor, voltage and current relate like this: $V_{rms} = I_{rms} R$. And for a capacitor, you have $V_{rms} = I_{rms} X_c$, where X_c is the capacitive reactance:

$$X_c = \frac{1}{2\pi fC}$$

So it shouldn't surprise you that for an inductor, you have another formula that relates *root-mean-square* voltage and current — the maximum voltage and current divided by the square root of 2 (for more on these terms, see the earlier section "Averaging out: Using root-mean-square current and voltage"):

$$V_{rms} = I_{rms} X_L$$

where X_L is the *inductive reactance* — that is, the effective resistance of the inductor: $X_L = 2\pi fL$.

Note that capacitive reactance gets big when the frequency of the applied voltage gets low, but inductive reactance gets big when the frequency gets big — opposite to capacitors. Why is this? It's because inductors act to oppose any change in the magnetic fields inside them. And the faster the applied voltage changes, the larger the change in flux divided by time — which means that the induced voltage can get really large when you go to a very high frequency.

Check out an example using inductive reactance. Say that you have an inductor with an inductance of $L = 3.60$ mH (millihenries), and you apply a voltage with a root-mean-square value of 25.0 volts across it at 100 hertz. What's the induced current in the inductor? Starting with $V_{rms} = I_{rms} X_L$, you see that

$$I_{rms} = \frac{V_{rms}}{X_L}$$

You know V_{rms}, so you need to figure out X_L:

$$X_L = 2\pi fL$$

Putting in the numbers you know gives you the following:

$$X_L = 2\pi(100 \text{ Hz})(3.60 \times 10^{-3} \text{ H}) \approx 2.26 \ \Omega$$

And plugging this into the equation for I_{rms} gives you the answer:

$$I_{rms} = \frac{25.0 \text{ V}}{2.26 \ \Omega} \approx 11.1 \text{ A}$$

So you'd get a pretty hefty 11-amp induced current.

Getting behind: Current lags voltage

How does the current in an inductor behave when you apply an alternating voltage? You can see the result in Figure 5-9, which graphs the current and the voltage in an inductor as a function of time. Note that here, current lags voltage — the opposite behavior from a capacitor, where current leads voltage. When current lags voltage, voltage reaches its peak before current does.

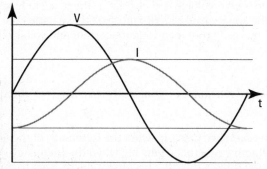

Figure 5-9:
Current lags
voltage in
an inductor.

Why does current lag voltage in an inductor? It's because of the following relation:

$$V = -L\frac{\Delta I}{\Delta t}$$

Note that this equation means that the voltage is greatest when the current is changing the fastest, because the voltage is directly proportional to the rate of change in current. So when the current is the steepest — when current is changing from negative to positive — it's changing the fastest, and voltage reaches its highest point. Conversely, when current is flat, it's not changing much at all, so the voltage goes to zero.

As you may expect, current lags voltage by exactly 90° — that is, $\pi/2$ in radians. So if the voltage is $V = V_0 \sin(2\pi ft)$, then current, which lags voltage by 90°, looks like this:

$$I = I_0 \sin\left(2\pi ft - \frac{\pi}{2}\right)$$

Using a little trig, this becomes the following:

$$I = -I_0 \cos(2\pi ft)$$

So that means that for an inductor, the power looks like this, because the current and voltage are 90° out of phase:

$$P = -I_o V_o \sin(2\pi ft) \cos(2\pi ft)$$

Note that, just as for a capacitor, the time average of $\sin(2\pi ft) \cos(2\pi ft)$ is zero, so the average power used by an inductor is zero:

$$P_{avg} = 0$$

The Current-Voltage Race: Putting It Together in Series RLC Circuits

Suppose you put a resistor, a capacitor, and an inductor together in the same circuit. The circuit in Figure 5-10 is called a *series RLC circuit* — *series* because all components are connected in series, one after the other (the same current has to flow through all of them) and *RLC* because it's a resistor-inductor-capacitor circuit (sometimes also called an *RCL circuit*). Note that the behavior of this circuit doesn't depend on the order of the circuit elements, so RLC or CLR would be just as good of a name.

Why doesn't the order of a resistor, inductor, and capacitor matter in a circuit? Consider the potential difference across each element: Across the resistor, the potential difference depends only on the current; across the inductor, it depends on only the rate of change of the current; and across the capacitor, it depends only on the sum of the current over time (that is, the charge). So the potential difference across each element depends only on the current, and the same current always flows through each in series, whatever order they're in.

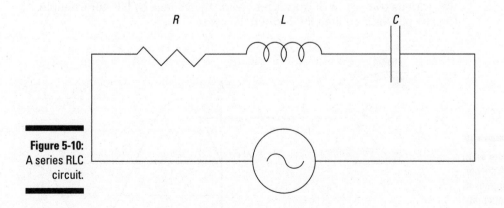

Figure 5-10:
A series RLC
circuit.

All the components are fighting each other over whether the voltage leads or lags — the capacitor wants the current to lead the voltage, the inductor wants the current to lag the voltage, and the resistor wants the voltage and the current to be in phase. Who wins? This section tells you where to place your bets.

Impedance: The combined effects of resistors, inductors, and capacitors

Earlier in this chapter, you see the relationship between root-mean-square current and voltage for the resistor, the inductor, and the capacitor. When you have a circuit combining various elements like resistors, capacitors, and inductors, then there's a similar relation for the circuit as a whole. The root-mean-square voltage across the circuit, per unit of root-mean-square current, is called the *impedance*.

Phasor diagrams: Pointing out alternating voltage and current

To tackle the problem of alternating voltages in an RLC circuit, you get a new tool: the *phasor diagram*. In this diagram, you represent the various alternating quantities as an arrow that rotates in time — you can see how this works in Figure 5-11:

- The arrow *(phasor)* on the left represents the alternating voltage V, with amplitude V_0. The length of the arrow is V_0.
- The arrow's angle from the horizontal, θ, is called the *phase*.

Now if you allow this arrow, initially horizontal, to rotate at a constant frequency f, then the phase is $\theta = 2\pi ft$. As you can see in Figure 5-11, if you project horizontally from the phasor at time t, then you get the value $V_0 \sin \theta$, which is just an alternating voltage. You can represent the current in the same way with its own arrow. If the current leads the voltage by 90°, for example, then its phasor is rotated 90° further clockwise.

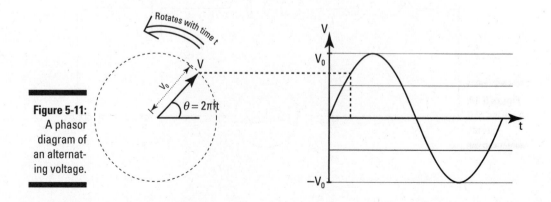

Figure 5-11: A phasor diagram of an alternating voltage.

Adding phasors and finding impedance

In Figure 5-12a, you can see three phasors representing the voltages across the resistor (V_R), inductor (V_L), and capacitor (V_C) in a circuit. In this figure, they're shown at time t, when the phase of the voltage across the resistor is $\theta = 2\pi ft$. Notice the voltage across the inductor leads the voltage across the resistor by 90°, and the capacitor lags by 90°.

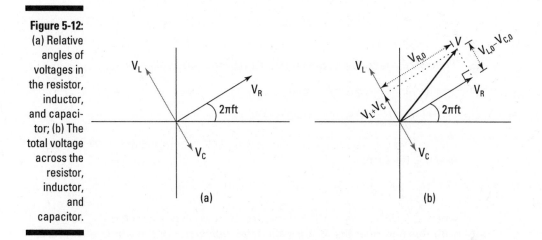

Figure 5-12:
(a) Relative angles of voltages in the resistor, inductor, and capacitor; (b) The total voltage across the resistor, inductor, and capacitor.

The *total potential difference* across all the circuit elements, V, is just the sum of the potential difference across each element. So to find V, add the phasors using vector addition (see Chapter 2 for info on adding vectors). Now, because V_L and V_C are always 180° out of phase, they simply point in opposite directions, so their sum is a new vector whose length is the difference in the amplitude of these two voltages.

The direction of this new phasor ($V_L + V_C$) is still 90° from V_R, because you've added two phasors that are both 90° from the phasor of V_R. To get the total sum, add V_R to this new phasor. You can see the sum of the voltages in Figure 5-12b. Because the phasors are at right angles, you can use the Pythagorean theorem to find the resulting length. The squared length of the sum of the potential differences is

$$V_0^2 = (V_{R,0})^2 + (V_{L,0} - V_{C,0})^2$$

where V_0, $V_{R,0}$, and $V_{C,0}$ are the amplitudes of the voltages.

Now if you use the relation between the amplitude V_0 and the root-mean-square voltage V_{rms}, you can use this to write the root-mean-square total voltage as

$$V_{rms}^2 = V_{R,rms}^2 + (V_{L,rms} - V_{C,rms})^2$$

where $V_{R,rms}$, $V_{L,rms}$, and $V_{C,rms}$ are root-mean-square voltages across the resistor, inductor, and capacitor respectively.

To simplify this equation, recognize that the following equations are true (note that because the current flows through all the components in series, only one current, I_{rms}, is in the circuit):

- $V_{R,rms} = I_{rms} R$
- $V_{C,rms} = I_{rms} X_C$
- $V_{L,rms} = I_{rms} X_L$

Therefore, you can put the equations together and solve for V_{rms}:

$$V_{rms}^2 = I_{rms}^2 [R^2 + (X_L - X_C)^2]$$
$$V_{rms} = I_{rms} [R^2 + (X_L - X_C)^2]^{1/2}$$

Now you're getting somewhere — you have V_{rms} in terms of I_{rms}. This equation has the form

$$V_{rms} = I_{rms} Z$$

where $Z = [R^2 + (X_L - X_C)^2]^{1/2}$. Very nice. Now you've connected V_{rms} to I_{rms} with this new quantity, Z. Z is called the *impedance* of the whole series RLC circuit, and it functions like the effective resistance of the whole RLC circuit.

Determining the amount of leading or lagging

For a series RLC circuit, $V_{rms} = I_{rms} Z$ (see the preceding section to find out where this equation comes from). That connects V_{rms} and I_{rms} in terms of their magnitude. But which leads — voltage or current? And by how much?

Look at a voltages-as-vectors graph. In Figure 5-13, I've added I (which is in phase with the voltage across the resistor, so it overlaps V_R) as a thick vector.

The question of whether voltage or current leads becomes, "What's the angle θ (as shown in the figure) between V and I?" Here's why:

- If that angle is positive, the net result of all three components is that the voltage leads the current.
- If that angle is negative, voltage lags the current.

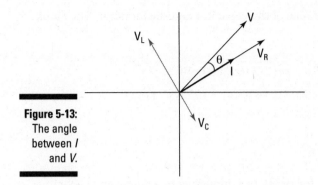

Figure 5-13:
The angle
between *I*
and *V.*

According to the figure, the tangent of this angle is

$$\tan\theta = \frac{V_{L,0} - V_{C,0}}{V_{R,0}}$$

$$= \frac{V_{L,rms} - V_{C,rms}}{V_{R,rms}}$$

$$= \frac{I_{rms}X_L - I_{rms}X_C}{I_{rms}R}$$

Note that in the second line, I've used the fact that the root-mean-square voltage is just the amplitude divided by the square root of 2; then I canceled the square root of 2 from the top and bottom of the fraction. Canceling out I_{rms} in the last line gives you

$$\tan\theta = \frac{X_L - X_C}{R}$$

So take the inverse tangent, $\tan^{-1}$, to find θ:

$$\theta = \tan^{-1}\left(\frac{X_L - X_C}{R}\right)$$

That's the angle by which voltage leads or lags the current across all three elements. If $\theta = 0°$, the voltage is in phase with the current — the effects of the inductor cancel out those of the capacitor. If it's positive, the inductor is winning; if it's negative, the capacitor is winning.

Finding the root-mean-square current

How about some numbers? Say that you have a circuit consisting of a 148-ohm resistor, a 1.50-microfarad capacitor, and a 35.7-millihenry inductor. The circuit is driven by an alternating voltage source with a root-mean-square voltage of 35.0 volts at 512 hertz. What is the root-mean-square current through the circuit, and by how much does the current lead or lag the voltage?

Start by getting the reactance of the capacitor and the inductor. They look like this:

- ✔ **Capacitor:** $X_C = \dfrac{1}{2\pi fC} = \dfrac{1}{2\pi(512 \text{ Hz})(1.50 \times 10^{-6} \text{ F})} \approx 207 \ \Omega$

- ✔ **Inductor:** $X_L = 2\pi fL = 2\pi(512 \text{ Hz})(35.7 \times 10^{-3} \text{ H}) \approx 115 \ \Omega$

The impedance is

$$Z = [R^2 + (X_L - X_C)^2]^{1/2}$$

Plugging in the numbers for resistance, inductive reactance, and capacitive reactance gives you the following:

$$Z = [148 \ \Omega^2 + (115 \ \Omega - 207 \ \Omega)^2]^{1/2} \approx 174 \ \Omega$$

And because $I_{rms} = \dfrac{V_{rms}}{Z}$, you get the following answer for the current:

$$I_{rms} = \frac{35.0 \text{ V}}{174 \ \Omega} = 0.201 \text{ A}$$

Quantifying the leading or lagging

Now, does the root-mean-square current lead or lag the voltage? In this example, the capacitive reactance (207 ohms) is greater than the inductive reactance (115 ohms), so you can say that the capacitor wins here and the voltage lags the current. But by how much?

Take the following equation:

$$\tan\theta = \frac{X_L - X_C}{R}$$

Just plug in the numbers. Because the resistance is 148 ohms, you find that

$$\tan\theta = \frac{115 \ \Omega - 207 \ \Omega}{148 \ \Omega} \approx -0.62$$

So take the inverse tangent to find the angle:

$$\theta = \tan^{-1}(-0.62) \approx -32°$$

And there you have it — voltage does indeed lag the current, just as in a capacitor.

Peak Experiences: Finding Maximum Current in a Series RLC Circuit

Earlier in this chapter, the resistor, capacitor, and inductor all have fixed values, as does the applied voltage. But all those things can vary: You can use electrical components that let you vary their resistance, their capacitance, their inductance, their voltage — even the frequency of that voltage. If you're going to vary anything in an RLC circuit, varying the frequency is the most common choice. This section tells you how to find the frequency at which you get the most current.

Canceling out reactance

When you let various quantities vary in an RLC circuit, the amount of current through the circuit changes. Because $V_{rms} = I_{rms} Z$, where $Z = [R^2 + (X_L - X_c)^2]^{1/2}$, you have the following:

$$I_{rms} = \frac{V_{rms}}{Z}$$

Note that the current through the circuit (I_{rms}) will reach a maximum when Z, the impedance, reaches a minimum — that is, when Z is at its smallest value. Because $Z = [R^2 + (X_L - X_c)^2]^{1/2}$, impedance will reach its minimum value when the inductive reactance equals the capacitive reactance:

$$X_L = X_c$$

At that point, $Z = R$.

Note that in this case, when the circuit is in resonance and the effects of the inductor and capacitor cancel each other out, the current and voltage are in phase.

Finding resonance frequency

The frequency at which the current reaches its maximum value is called the *resonance frequency*. At the resonance frequency, the effects of the capacitor and the inductor cancel out, leaving the resistor as the only effective element in the circuit.

Resonance: Getting big vibrations

Resonance is not just a feature of electrical circuits; it's a general feature of oscillating systems — pendulums and even bridges and skyscrapers can experience it. The oscillating system may wobble with a particular amplitude when driven at a particular frequency, such as when you apply an AC voltage of frequency *f* to your circuit or when an earthquake shakes a skyscraper.

If you want to make things wobble the most, it's not a case of driving at the highest frequency that you can. The system likes to wobble at a certain natural frequency, and if you drive it at this frequency, then you get the biggest response — this is the *resonance frequency*. There's a particular frequency in the circuit that gives the greatest amplitude current if you apply the voltage at that frequency. (By the way, anyone designing a skyscraper will make sure that its resonance frequency is different from the frequency at which earthquakes shake!)

What is the resonance frequency for any given RLC circuit? You know that

$$X_C = \frac{1}{2\pi f C} \text{ and } X_C = 2\pi f L.$$

At the resonance frequency f_{res}, the inductive reactance and capacitive reactance are equal, so the following equation holds:

$$2\pi f_{res} L = \frac{1}{2\pi f_{res} C}$$

Rearranging this equation and solving for frequency gives you

$$f_{res}^2 = \frac{1}{(2\pi)^2 LC}$$

$$f_{res} = \frac{1}{2\pi (LC)^{1/2}}$$

And there you have it — that's the frequency at which the current reaches its maximum value for any given L and C values.

Semiconductors and Diodes: Limiting Current Direction

One of the great leaps of the technological age happened when people started combining the resistor and the capacitor with some new circuit elements made from materials that were semiconductors. The combination was an extremely powerful one. These circuits, combining resistors, capacitors, and semiconductor devices, eventually became miniaturized into *integrated circuits,* or *microchips,* which form the basis for many devices that have changed the way people live — most notably the computer. (So the next time someone complains you're spending too much time on the computer, tell them you're doing physics.)

In this section, I first introduce semiconductors so you understand their special properties. Then I introduce an example of a circuit element made from them: the *diode*. This simple device allows current to pass through it in one direction only — it's effectively a one-way valve for electrical current.

The straight dope: Making semiconductors

Normal silicon (Si) has a crystalline structure, with four electrons from each atom taking part in bonding each atom to its neighbors. Those electrons are in the outermost orbits of the silicon atom, and because they're important in creating the crystalline structure, they're not available to conduct electricity — hence, normal silicon is an insulator.

But by being clever, you can introduce small amounts of impurities (such as one part in a million) that give the silicon conducting properties. Here are two types of semiconductors you can create:

✔ **N-type semiconductors:** Adding some phosphorus (P) atoms allows the silicon to conduct electricity. Phosphorus has five electrons in its outermost orbit, so when you *dope* silicon with phosphorus, the phosphorus atoms join the silicon crystal structure, which binds each atom to its neighbors using four electrons. That means that there's one electron from the phosphorus left over — and that electron is free to roam.

The resulting doped silicon is called an *n-type semiconductor,* because the charges that carry current in it — the electrons contributed by the phosphorus — are negative.

✔ **P-type semiconductors:** On the other hand, you can dope silicon with other elements, such as boron (B), which has only three outer electrons per atom. When the boron binds to the silicon-crystal structure, one electron is missing, so there's a "hole" in the number of electrons.

That hole can move from atom to atom — and each hole produces a positive charge, because it's formed from a deficit of electrons. Because the holes (that is, the localized places where you have a missing electron) can move throughout the semiconductor, the charge-carriers in this kind of doped silicon are positive. When you have a material with mobile holes, it's called a *p-type* semiconductor, because the free charge-carriers are positive.

That's the whole charm of semiconductors — in addition to negatively charged carriers (electrons), you can also have positively charged carriers (the holes).

One-way current: Creating diodes

You can create *diodes* — one-way current valves — by putting some *p*-type semiconductor next to some *n*-type semiconductor (see the preceding section for info on types of semiconductors). In the case at the top of Figure 5-14, voltage is applied with the positive voltage connected to the *p*-type semiconductor, and negative voltage is connected to the *n*-type semiconductor.

In this case, charge flows freely across the junction between the *p*-type and *n*-type semiconductors, because the positive holes on the left are repelled from the positive terminal and travel to the right, and the electrons on the right are repelled by the negative terminal, so they travel to the left. The holes and electrons meet at the junction, and the electrons fill the holes — so current can flow. The negative terminal provides more electrons for this process, and the positive terminal removes them, creating more holes.

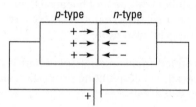

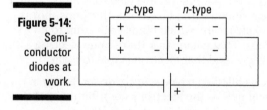

Figure 5-14:
Semi-
conductor
diodes at
work.

On the other hand, if you reverse the terminals of the battery, no current will flow through the diode, as the bottom of Figure 5-14 shows. That's because in this case, the battery drives the mobile charge-carriers away from the junction. As you can see in the figure, the positive holes travel to the left in the *p*-type semiconductor — away from the junction — and the electrons in the *n*-type semiconductor travel to the right, also away from the junction.

What's left at the junction are the immobile negative charges in the *p*-type material and the immobile positive charges in the *n*-type material. Those charges don't move, so they set up an electric field that counteracts the electric field set up by the battery — with the net result that all current stops.

Part III
Catching On to Waves: The Sound and Light Kinds

The 5th Wave   By Rich Tennant

"And that's the Doppler Effect."

In this part . . .

In this part, you take a look at waves, specifically sound and light waves. You get the lowdown on sound waves and then spend a few chapters on how light waves work, including what happens when they hit mirrors, bend through lenses and diamonds, and pass through slits. Light wave behavior is one of the favorite topics of physicists, and in this part, you see why.

Chapter 6

Exploring Waves

*W*aves are all around you — water waves, sound waves, light waves, even waves in jump ropes. (Do the waves in that starlet's hair count? Not in this chapter.) Waves are such a huge topic in Physics II that I cover them in detail in the next five chapters. In fact, even matter travels in waves and is subject to the same kinds of effects as light waves, including reflection (see Chapter 12 for details on this surprising behavior).

In this chapter, you investigate just what waves are and how they work — and how to describe them mathematically (physicists love describing things mathematically). You work with formulas and get to do a little graphing, too. I wrap up by describing some typical wave behavior. Later, in Chapters 7 through 11, you work with specific types of waves: sound and light.

Energy Travels: Doing the Wave

Understanding waves begins with being able to recognize their characteristics. Here are a few key features of waves that you can discover just from watching water waves:

✔ **A wave is a traveling disturbance.** Waves don't occur when a surface such as water is calm. Suppose you and some friends are in a sailboat on a lake when a motorboat roars past, sending your boat bobbing. First, you notice that the surface of the lake is now filled with waves and ripples. The water was disturbed by the motorboat, and that disturbance is being sent all around the lake. When a lake is calm, you don't have any waves; when a lake is disturbed, you have waves. So something must disturb the water in order to create water waves. The thing that's disturbed by a wave is called the *medium*.

✔ **A wave transfers energy.** All waves transfer energy. In fact, waves are one of the primary means of getting energy from Point A to Point B. Continuing with the earlier example, you realize your sailboat is being lifted up and down in the wake of the motorboat. Lifting the boat takes energy — elevating the boat adds potential energy to it. The humps of water in the waves surrounding you all have potential and kinetic energy.

✔ **A wave doesn't cause bulk transport of the underlying medium (if there is an underlying medium).** As a wave travels, the medium wobbles, or *oscillates,* about its undisturbed position, but it doesn't shift on the whole — this is what I mean by "no bulk transport." Each part of the medium oscillates about its resting state without changing on average.

For example, suppose you notice a leaf floating on the lake, going up and down with each passing wave. Even though the waves look like they're traveling away from your boat, the leaf isn't moving anywhere except up and down. That's because the water isn't really traveling across the lake — the wave is. The wave seems to move on to the next patch of water, then the next, and so on, without making any one part of the water travel across the lake. That is, there's no *bulk* movement of the water. No mass of water is moving across the lake; each wave just moves each successive region of water up and down as it passes.

Waves — these traveling disturbances carrying energy — come in two types: transverse and longitudinal. The kind depends on which direction the energy disturbance is traveling. This section takes a look at both wave types.

Up and down: Transverse waves

A *transverse wave* moves up and down, creating peaks of movement. The motion of this type of wave disturbance is perpendicular to the direction the wave is moving in. If you've ever had a vacuum cord get stuck while you were vacuuming and yanked on the cord to dislodge it, you saw a transverse wave in action. When you whipped the cord up and down to free it, waves traveled up and down the cord a little something like Figure 6-1.

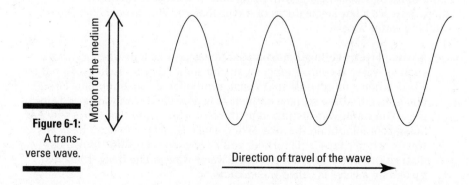

Figure 6-1: A transverse wave.

Motion of the medium

Direction of travel of the wave

Back and forth: Longitudinal waves

In *longitudinal waves,* the motion of the wave disturbance is parallel to the direction the wave is traveling in. As the different parts of the medium wobble back and forth in the direction of the wave's travel, they cyclically squash and stretch along the wave. A physicist may call a squashing of the medium *compression* and the stretching *decompression.*

This kind of wave can travel only in a medium that's capable of being stretched and squashed — that is, an *elastic* medium. For example, a spring can support compression and decompression down its length, but a string can't. Figure 6-2 depicts a longitudinal wave traveling in repeating cycles of compression and decompression, or *pulses.*

Most objects are elastic to some extent, so you can send pulses through them. Pulses in the air are referred to as *sound,* which carries the energy from far-off disturbances to your ears. I discuss sound in Chapter 7.

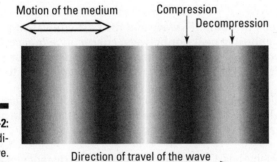

Motion of the medium

Compression

Decompression

Figure 6-2: A longitudinal wave.

Direction of travel of the wave

Wave Properties: Understanding What Makes Waves Tick

All waves, no matter which direction they're traveling in, have specific parts and properties, such as periods and frequency. In this section, you discover the details of a wave's basic parts and properties. You also see how all the parts of a wave relate mathematically, as well as what a wave looks like in graph form.

Examining the parts of a wave

To understand waves, you need to have a good grip on the terminology. (How else can you discuss waves with your fellow physicists-in-training?) Take a look at Figure 6-3, which lists some important parts of a wave. The subsections that follow delve into these parts in greater detail.

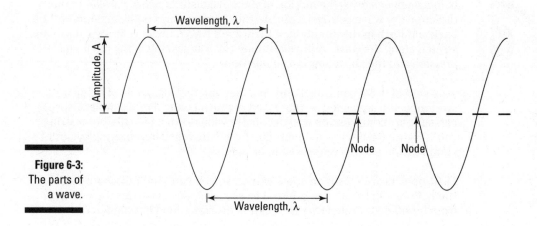

Figure 6-3:
The parts of
a wave.

Wavelength

The distance between one point of a wave and the next equivalent point — such as between neighboring peaks or between consecutive *troughs* (the lowest points on a wave) — is known as the wave's *wavelength*. For a longitudinal wave, the wavelength is the distance from one compression to the next.

Nodes are specific locations where a wave crosses the axis; there are always two nodes per wavelength. The parts of the medium that are at the nodes of the wave are in their resting, undisturbed positions.

The symbol for wavelength is λ. You usually measure the distance of a wavelength in meters — unless you're dealing with light waves, which are typically measured in a much smaller unit called *nanometers* (nm), which are billionths of a meter.

Amplitude

A wave is a traveling disturbance, and the wave's *amplitude* tells you how big that disturbance is. Amplitude represents different things depending on whether you're working with a transverse wave or a longitudinal wave. The amplitude of a transverse wave is a measure of the distance from the axis to a peak, or from the axis to a trough (that should be the same distance). In other words, amplitude is a measure of how high a wave is (see Figure 6-3). Generally, the amplitude of a wave is half of the peak-to-trough distance.

For longitudinal waves, such as sound waves, amplitude corresponds to the pressure in each pulse. I explain the amplitude of sound waves in Chapter 7.

The symbol for amplitude is *A,* but the units of measurement for amplitude vary depending on which kind of wave you're dealing with. For example, the amplitude of a water wave on the surface of a lake is measured in units of distance (such as meters or feet) because you're trying to find out how high the wave is. The amplitude of a light wave, on the other hand, which alternates between magnetic and electric fields, can be measured in teslas and volts per meter (although the amplitude is truly tiny amounts of both).

Periods and cycles

Waves are *periodic,* alternating and repeating in a certain amount of time, as you can see in Figure 6-3. If you go from one part of a wave to the same part again — like from peak to peak in a transverse wave or compression to compression in a longitudinal wave — you've gone through one *cycle.* In other words, if you see five peaks or compressions go past, you know that five wave cycles have been completed.

The time it takes to complete a cycle is referred to as the wave's *period.* So if you see a peak of a transverse wave, wait a moment, and see another peak, you know that one period has passed. You measure periods (symbol *T*) in seconds.

Frequency

Frequency measures the number of times something happens per second. Wave frequency is measured in cycles per second. And because cycles are just numbers, that means the unit for frequency is s^{-1}. Of course, s^{-1} goes by another, more common name: *hertz* (symbol Hz).

The symbol for frequency is *f.* To calculate frequency, just take 1 over the period *(T)*, like so:

$$f = \frac{1}{T}$$

So, for example, a wave that has a period of $\frac{1}{100}$ seconds has a frequency of 100 cycles per second, or 100 Hz.

Relating the parts of a wave mathematically

Knowing the parts and properties of waves is all well and good, but you also need to be able to do something with them. That's where the math comes in. By applying a little math to what you know about waves, you're in a position to say more about them. For instance, you can tell someone how fast a particular wave travels, or you can figure out the wavelength. This section shows you how.

Getting a general formula for wave speed

Speed is the distance traveled divided by the time it took to go that distance, so the speed of a wave is simply the distance that a peak travels divided by the time it took to do so. In other words, you divide the wavelength by the period like this:

$$v = \frac{\lambda}{T}$$

Because the frequency, *f,* is $1/T$, you can write the basic equation for calculating the wave speed as

$$v = \lambda f$$

A short message from our sponsors: Calculating wavelength of a radio signal

Try putting some numbers in the general wave speed formula. Say that you're listening to a radio station, 1230 AM on your dial. What's the wavelength of that radio signal?

The frequency of the wave is 1230, but 1230 what? AM frequencies are measured in kHz (kilohertz), so that's a frequency of $1{,}230 \times 10^3$ Hz, or 1.23×10^6 Hz.

Because $v = \lambda f$, you can rearrange the formula to solve for wavelength:

$$\lambda = \frac{v}{f}$$

All you need now is the speed of the radio signal. Radio signals travel at the speed of light ($v \approx 3.00 \times 10^8$ meters per second), so plug in the numbers and solve:

$$\lambda = \frac{3.00 \times 10^8 \, \text{m/s}}{1.23 \times 10^6 \, \text{Hz}} \approx 244 \text{ m}$$

So the wavelength is about 244 meters, or 800 feet. The next time you're listening in on a frequency of 1230 kHz, you can say you're listing in on a wavelength of 800 feet. Or if you really want to blow your mind, think of the radio signal as a wavelength 800 feet long coming at you 1.23 million times a second. Whoa!

A tense situation: Figuring out the speed of a transverse wave

Sometimes, you can say more than just $v = \lambda/T$ — you can figure out what the wave speed is for a given setup using properties of the system itself. For example, if you have a string under tension, you can calculate the speed of waves in the string given only the force of tension, the mass of the string, and its length.

Actually, you don't even need to know the mass and the length of the string — you just need to know the mass per unit length, μ, which is

$$\mu = \frac{m}{L}$$

where m is the mass in kilograms and L is the length in meters.

At tension F (where F stands for force), the speed of transverse waves in the string turns out to be

$$v = \left(\frac{F}{\mu} \right)^{1/2}$$

That makes sense — the stronger the tension (the larger F is), the faster the waves go, and the heavier the string (the larger μ is), the slower the waves go.

Say that you have string that's 20 grams per meter, and it's under a tension of 200 newtons. How fast does a transverse wave travel in the string if you pluck it? You know that $v = (F/\mu)^{1/2}$, so plug in the numbers (after converting to kilograms) and solve:

$$v = \left(\frac{200 \text{ N}}{0.020 \text{ kg/m}} \right)^{1/2}$$

$$= \left(1.0 \times 10^4 \text{ m}^2/\text{s}^2 \right)^{1/2} = 100 \text{ m/s}$$

So the speed of the transverse wave is 100 meters per second.

Watching for the sine: Graphs of waves

Graphing a wave gives you an idea of how a wave changes over time. When you graph a wave, whether it's transverse or longitudinal, you're really plotting the magnitude of the disturbance. That may be the magnitude of the string displacement or the magnitude of the pulsing water pressure. Because you're just graphing magnitude, you can graph both transverse and longitudinal waves as sine waves.

Consider the correlation between sine waves and transverse waves: Transverse waves (the kind you create when you whip a string up and down) look just like sine waves. There's a good reason for that — they are sine waves!

Longitudinal waves are pulses in the direction of travel, which means they don't look like sine waves. But if you graph the magnitude of the disturbance along a longitudinal wave — pulses and all — you find that a longitudinal wave from a continuous source looks like a sine wave.

Picture a long succession of longitudinal waves traveling through water. Each pulse corresponds to a peak of a sine wave, and the space between pulses corresponds to a trough. So in this case, when you plot the pressure in the wave due to a passing longitudinal wave, you actually get a sine wave (if, of course, the wave source creates normal longitudinal waves).

In this section, I explain how to graph a sine wave that accurately describes a physical wave.

Creating a basic sine wave

So what exactly do you need to graph a real-world wave? First, you have to know what your axes are. Because you're measuring the magnitude of the disturbance created by the wave, your vertical axis is displacement (y). And because you want to know how long that disturbance occurs, your horizontal axis is time (t).

You want to complete one cycle of the sine wave you're drawing in one cycle of the actual wave. A single cycle of a wave takes place in one period, and a single cycle of a sine wave takes place in 2π radians. That means that in one period, you want the sine wave to go through 2π radians, as Figure 6-4 shows. You can use this expression for the sine wave:

$$y = \sin\left(\frac{2\pi t}{T}\right)$$

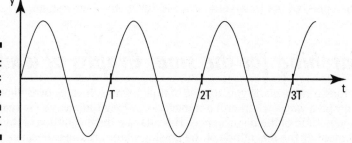

Figure 6-4: The basic sine wave with period T.

Note that when $t = 0$, $y = 0$. And when $t = T$, you have $y = \sin(2\pi)$, which equals 0.

You can get frequency into the equation with a little substitution, because you can relate a wave's period and frequency like so (as I explain in the earlier section "Frequency"):

$$f = \frac{1}{T}$$

Substitute *f* into the expression for the sine wave to write the expression as

$$y = \sin(2\pi f t)$$

Adjusting the equation to represent a real-world wave

The wave equation $y = \sin(2\pi f t)$ is fine, but you probably need to stretch or shift your graph so it accurately depicts your real-world wave. Otherwise, the graph doesn't give you any information on the strength of the wave or where it was in its cycle when you started taking measurements.

You want your graphed wave to have its own amplitude, *A,* to show how big the disturbance is — a bit tricky to manage because sine waves oscillate between –1 and 1. Multiply the sine wave by *A* to get the following:

$$y = A \sin(2\pi f t)$$

Of course, the wave's displacement doesn't have to be at 0 when *t* = 0. In Figure 6-5, the wave starts off at a nonzero value when *t* = 0, so you need to adjust your wave expression to take this shift into account. Good news: You can adjust the sine's *argument* (the value you're taking the sine of) by an angle, called the *phase angle,* to make your graphed wave match the behavior of the actual wave.

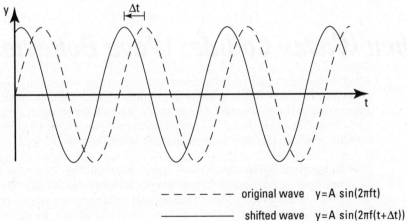

Figure 6-5:
An offset
wave.

– – – – – – original wave   y=A sin(2πft)

———————— shifted wave   y=A sin(2πf(t+Δt))

Here's how to add a phase angle to the expression for a wave (note that θ can be positive or negative):

$$y = A \sin(2\pi f t + \theta)$$

You can also write this equation in terms of a time shift, Δt. Here's how:

$$y = A \sin(2\pi f(t+\Delta t))$$

This means that the original wave is shifted in time so that a peak, originally happening at time t, now happens Δt earlier in the shifted wave.

Note that if $\theta = \pi/2$, you get a cosine wave:

$$y = A \, \sin\left(2\pi ft + \frac{\pi}{2}\right) = A \, \cos\left(2\pi ft\right)$$

If you shift the wave by one whole period, it looks exactly like the original. You can see this is true because you know from basic trig that $\sin(x + 2\pi) = \sin(x)$. So if you shift the wave by $\Delta t = T$, the wave becomes

$$y = A \sin(2\pi f(t + \Delta t))$$

$$= A \sin(2\pi ft + 2\pi fT)$$

$$= A \sin(2\pi ft + 2\pi)$$

$$= A \sin(2\pi ft)$$

and you have your original wave back again.

When Waves Collide: Wave Behavior

Most waves can't just travel forever without hitting something — some object, or maybe another wave — and that's what makes wave behavior interesting in the real world. For example, when light waves travel through a glass lens, the waves bend, so people can create eyeglasses and telescopes and binoculars. Here are some important wave behaviors:

- ✔ **Refraction:** When waves enter a new material, they can alter their behavior — change their wavelength, for example, or alter their direction. Light waves do this in lenses and prisms, water waves do this in the shallows, and sound waves do this when traveling from air to glass. This process is called wave *refraction*, and I cover refraction of light waves in Chapter 9.

- ✔ **Reflection:** When waves hit something, such as when light waves hit a mirror, they can bounce off, a process known as *reflection*. Sound waves can reflect off walls, radio waves can reflect off layers of the atmosphere, TV signals can reflect off buildings, and so on. You can find lots more on reflection in Chapter 10.

✔ **Interference:** Waves can also hit each other, and when they do, they interfere — and the resulting process is called *interference*. For example, you may have seen the ripples from two stones thrown into a lake overlap — and the result is called an *interference pattern*. The waves' amplitudes can add to each other or cancel each other out. You can find a great deal on interference in light waves in Chapter 11.

Chapter 7

Now Hear This: The Word on Sound

Sound is all around you — the sound of talking, the sound of leaves rustling, the sound of traffic, even *The Sound of Music*. Sound travels in perfect longitudinal waves (that is, the wave's disturbance travels in the same direction as the wave; see Chapter 6 for details). As such, sound waves are a fit topic for physicists.

You get the lowdown on sound in this chapter — how it works, what it can do, and what it can't do — starting with a look at sound waves as vibrations. You then explore ideas such as the speed of sound, loudness, echoes, and more.

Vibrating Just to Be Heard: Sound Waves as Vibrations

Sound is a vibration in the medium through which the sound is traveling — air, water, metal, or even stone. But it's not just any vibration; it's actually a vibration *caused by* a vibration. A vibrating object makes the air surrounding it vibrate, too, and those vibrations travel away from the vibrating object through the air.

Say you're dealing with the *diaphragm* in a loudspeaker (that's the part that vibrates) and it's vibrating furiously, pumping out some loud music. Each time the diaphragm pushes against the air, it compresses the air near it. That creates a *condensation* in the air. This kind of condensation is a small, high-pressure region in the air — a local pulse. As soon as the speaker diaphragm creates the condensation, that condensation starts traveling off into the air.

Conversely, when the diaphragm springs back, that movement creates a small low-pressure region, known as a *rarefaction,* in the air around the diaphragm. Just as with the condensation, as soon as a rarefaction is created, it starts traveling away from the loudspeaker through the air. Those alternating condensations and rarefactions travel through the air as a longitudinal wave — much like the pulses you can send through a spring when you rapidly compress and decompress one end of it.

So there you have it: Sound is really a longitudinal wave that travels through the air in a series of condensations and rarefactions — that is, *pulses.* In Figure 7-1, I've magnified the column of air that shoots out from the loudspeaker so the air molecules are actually visible. Notice how the air molecules are close together in the condensations and spread out in the rarefactions.

Figure 7-1:
Pulses in
a random
sound wave.

Normal music is made up of many different sound waves, so the pulses you see coming from the loudspeaker have different amplitudes and different frequencies. When these waves enter your ear, the oscillation of the air causes your eardrum to vibrate, and your brain interprets these sounds as having pitch and loudness. Here's how a sound wave's amplitude and frequency affect what you hear:

- ✔ **Amplitude:** If a sound wave entering your ear has a large amplitude, then you hear a louder sound.

- ✔ **Frequency:** If a sound wave entering your ear has a high frequency, then you hear a high-pitched sound. But this can vary from person to person because the sensitivity of different people's ears to different frequencies of sounds varies.

The human ear can hear a wide range of sound frequencies. Newborns, for example, can hear from 20 hertz (Hz) up to an astounding 20,000 Hz. As you age, you can't hear the upper range quite so well. An adult, for example, may hear only up to 14,000 Hz. Sound with a frequency higher than 20,000 Hz is called *ultrasonic,* and sound with a frequency lower than 20 Hz is called *infrasonic.*

When you have sound that comes out of a loudspeaker in a pure, unwavering tone, the condensations and rarefactions all have the same strength, and they're all evenly spaced, as in Figure 7-2. The figure shows the waves of condensation and rarefaction of the molecules (not actual size!). You can see that as the molecules are displaced back and forth, they go through cycles of high and low pressure. Where the molecules are squeezed together in a condensation, the pressure is high, and when they're stretched apart in a rarefaction, the pressure is low — I've plotted the wave's fluctuation in the following graph.

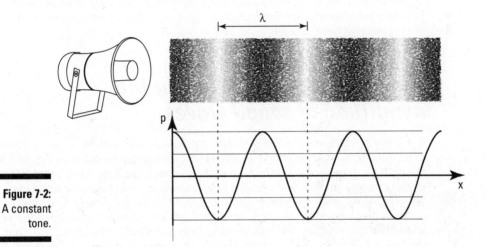

Figure 7-2:
A constant
tone.

When you have a single tone coming from a loudspeaker, you can speak of the wavelength of the sound, λ, and its frequency, f. Regular sound waves, like the one in Figure 7-2, have so many cycles per second, which is their frequency.

Cranking Up the Volume: Pressure, Power, and Intensity

The loudness, or volume, at which you hear a sound is a direct result of the sound wave's *amplitude* — that is, the amount of pressure in each pulse in a sound wave. The greater the pressure amplitude, the greater the volume.

Volume is really a subjective measure; a sound may seem louder to one person than to another, based on how good his or her hearing is. But in physics, you use objective measures, such as pressure amplitude and sound intensity, to talk about the sonic boom that rattled your windows or the rock concert that still has your ears ringing.

Amplitude and sound intensity are related. Here's how: Making a sound wave takes energy, and making a continuous wave takes a flow of energy over time: *power*. As a wave propagates and spreads out in the surrounding space, this power is spread over a larger area, so sound can become weaker with distance. The amount of power flowing through a unit area is its *intensity*. Lower intensity causes less energy to enter your ear every second — and with less sound power entering your ear, the wave has a smaller amplitude, because making a wave with smaller amplitude takes less power. That's why sounds become quieter with distance.

In this section, I discuss the amplitude, power, and intensity of sound waves. Intensity is related to decibels, a way to compare sounds objectively, so I cover decibels here as well.

Under pressure: Measuring the amplitude of sound waves

If you wanted to measure the pressure amplitude of sound waves (or if some crazy professor said you had to), you could start with the setup in Figure 7-3. There, a loudspeaker is sending a pure-tone sound through a tube that has many pressure meters at the top of it. (A *pure-tone sound* is made up of just one frequency, so it's a *monofrequency sound*.) Using this setup, you can measure the amplitude of the traveling sound wave by photographing the settings of all the pressure meters at once. This snapshot can also tell you that the pressure in the whole wave forms a sine wave that's traveling to the right.

Figure 7-3:
Measuring
sound pres-
sure in a
sound wave.

Suppose the loudspeaker in Figure 7-3 is set to create a monofrequency wave at about the volume of human speech, and you need to find the maximum amplitude. Pressure is measured in *pascals* (Pa), and it takes 1.01×10^5 Pa to make up the pressure of the atmosphere at sea level. The maximum pressure amplitude of human speech is about 3.0×10^{-2} Pa, or an amazing 3.0×10^{-7} atmospheres! That's how sensitive the human ear is. Human speech, which can sound very loud, is actually made up of very weak pulses of air. So the pressure amplitude of a sound wave of human speech is relatively small.

Even though sound is a longitudinal wave, you can graph its pressure amplitude as a sine wave, because you're measuring the displacement of air (it's just like the amplitude of a transverse wave in a string, because what you measure there is the actual displacement of the string). For a sound wave, condensations form the peaks of the sine wave and rarefactions form the troughs.

Introducing sound intensity

Sounds waves transfer a disturbance in a medium from the source to an observer. That means energy is transferred from the source to some target. Leaves rustling in the street transfer a relatively small amount of energy, but some sounds are powerful enough that they can cause damage. Sonic booms, for example, are strong enough to break windows.

So how much energy is transferred by a sound wave in a given amount of time? That's a measure of *power,* which is measured in watts (abbreviated W). Power is just energy divided by time:

$$P = \frac{E}{t}$$

In fact, what's usually measured is the power per unit area some distance from the sound source, as Figure 7-4 shows. This quantity, power divided by area, is the sound wave's *intensity.* Sound intensity is measured in watts per meter, and the equation for finding it is as follows:

$$I = \frac{P}{A}$$

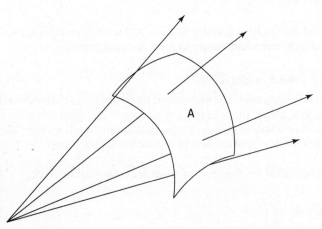

Figure 7-4:
Sound intensity is the power of a sound wave divided by the area.

In this section, you calculate sound intensity and see how it relates to decibels.

Sound intensity in terms of total power of a sound wave

The intensity of a sound wave differs depending on how far away you are from a sound source. That's because sound expands in a sphere from a sound source, and the power of a sound wave is distributed over the whole area of that sphere. The following equation shows how the surface area of a sphere (A) grows as you get farther away from the sound source — that is, as the radius increases:

$$A = 4\pi r^2$$

where r is your distance from the sound source.

If you know the total power of a sound wave as it comes out of the source, P_{total}, and you know that sound wave is allowed to expand in a sphere, you can write the intensity as a function of r like so:

$$I = \frac{P_{total}}{4\pi r^2}$$

Thus, the intensity of a sound wave drops off by a factor of 4 (or 2^2) every time you double your distance from the sound source.

For example, say you have a sound source that pumps out 3.8×10^{-5} watts of sound power. What's the sound intensity 1 meter from the sound source? Well, the total power of the sound energy the source sends out is 3.8×10^{-5} watts. At $r = 1.0$ meters, you have this:

$$I = \frac{P_{total}}{4\pi r^2}$$
$$= \frac{3.8 \times 10^{-5} \text{ W}}{4\pi (1.0 \text{ m})^2} \approx 3.0 \times 10^{-6} \text{ W/m}^2$$

So the sound intensity at 1 meter is 3.0×10^{-6} watts per square meter. That's the approximate sound intensity of human conversation.

Measuring sound in decibels

Decibels are a comparison of one sound intensity to a reference intensity on a logarithmic scale. In plain English, that means decibels tell you how much louder or softer a sound is than a standard sound, such as the threshold of hearing (that's the reference sound physicists usually use).

Here's the equation for decibels of a particular sound intensity:

$$\beta = 10 \ \log\left(\frac{I}{I_o}\right)$$

where *log* refers to the logarithm to the base 10 (it's on your calculator); I_o refers to the reference sound you're measuring against (usually the threshold of hearing, 1.0×10^{-12} W/m²); and I is the sound intensity you're measuring. The abbreviation for decibels is dB.

How about some representative numbers here? Table 7-1 lists some common decibel measurements from 1 meter away from the source, comparing the sound to the threshold of human hearing.

Table 7-1	Intensity and Decibels of Common Sounds	
Sound	*Intensity*	*Decibels*
Threshold of hearing	1.0×10^{-12} W/m²	0 dB
Leaves rustling	1.0×10^{-11} W/m²	10 dB
Whisper	1.0×10^{-10} W/m²	20 dB
Normal conversation	3.2×10^{-6} W/m²	65 dB
Car with no muffler	3.2×10^{-2} W/m²	100 dB

Say you have a gasoline-powered lawn mower that sounds especially loud, and you want to find out just how loud it is. You measure the sound intensity at 1 meter from the lawn mower as 6.9×10^{-2} W/m². How many decibels is that compared to the threshold of hearing?

The threshold of human hearing has a sound intensity of 1.0×10^{-12} W/m², so when you plug that into the $\beta = 10 \log(I/I_o)$ formula, you have

$$\beta = 10 \ \log\left(\frac{I}{I_o}\right)$$

$$\beta = 10 \ \log\left(\frac{6.9 \times 10^{-2} \ \text{W/m}^2}{1.0 \times 10^{-12} \ \text{W/m}^2}\right)$$

$$\beta \approx 108 \ \text{dB}$$

Your lawn mower generates about 108 dB at a distance of 1 meter away from the sound source. Hmm. Maybe you should start wearing earplugs when you use it!

Calculating the Speed of Sound

Sound traveling through the air moves pretty fast, but it can move even faster depending on which medium it's moving through (another gas, a liquid, or a solid). Of course, the only way to really know how fast it's traveling is to calculate its speed.

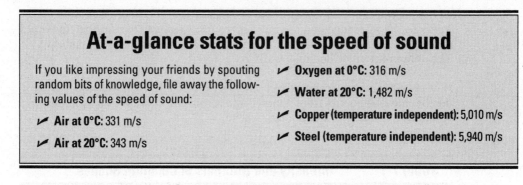

At-a-glance stats for the speed of sound

If you like impressing your friends by spouting random bits of knowledge, file away the following values of the speed of sound:

✔ **Air at 0°C:** 331 m/s

✔ **Air at 20°C:** 343 m/s

✔ **Oxygen at 0°C:** 316 m/s

✔ **Water at 20°C:** 1,482 m/s

✔ **Copper (temperature independent):** 5,010 m/s

✔ **Steel (temperature independent):** 5,940 m/s

The speed of a wave is frequency multiplied by wavelength, which looks like this:

$$v = \lambda f$$

However, that basic equation doesn't help you much because the speed of sound can vary depending on the temperature of the medium. But never fear — in this section, I introduce some speed-of-sound formulas that account for both temperature and medium. (And for some real-world values, check out the nearby sidebar titled "At-a-glance stats for the speed of sound.")

Fast: The speed of sound in gases

The speed of sound is lowest when it's traveling through a gas. To calculate the speed of sound in an *ideal gas* (which approximates air given the temperature of that gas), you rely on an equation that may look familiar to you from Physics I:

$$v = \left(\frac{\gamma k T}{m} \right)^{1/2}$$

Here's what the variables represent:

✔ γ is the *adiabatic constant,* and it's equivalent to C_p/C_v, the ratio of the specific heat capacity at constant pressure to the specific heat capacity at constant volume; for air, γ is 1.40.

✔ k is the Boltzman constant from thermodynamics (1.38×10^{-23} kg·m²s⁻²K⁻¹, or J/K).

✔ T is the temperature of the ideal gas according to the Kelvin scale.

✔ m is the mass of a single molecule in kilograms.

Okay, time to put this equation to work! Go ahead and assume you have in your hands a camera whose rangefinder uses sound to find the distance to the subject. You've just taken a photograph of a fellow physicist, and being a physicist, your friend immediately wants to know the distance between the two of you. Checking your camera, you can see that the rangefinder sent a pulse of sound out that bounced off your friend and came back to the camera in 4.00×10^{-2} seconds. Your handy pocket thermometer tells you that the temperature of the air is 23°C. So just how far away is your friend, assuming you can treat air as an ideal gas?

First, you need to convert that temperature to kelvins by adding 273 to the Celsius temperature, which looks like this:

23°C + 273 K = 296 K

So the ideal temperature is 296 kelvins. Great. Now you can use the speed-of-sound equation for gases:

$$v = \left(\frac{\gamma kT}{m} \right)^{1/2}$$

Notice that in addition to the temperature, you also need the mass m of a single molecule of air in kilograms. You just happen to remember that the mass of air is 28.9×10^{-3} kg/mole (a *mole* is 22.4 liters of ideal gas). So the mass of one air molecule is the mass of a mole divided by the number of molecules in a mole (Avogadro's number).

$$m = \frac{28.9 \times 10^{-3} \text{ kg/mole}}{6.022 \times 10^{23} \text{ molecules/mole}} \approx 4.80 \times 10^{-26} \text{ kg}$$

Bet you always wanted to know that! Right, moving on. For air, γ is 1.40, so this equation allows you to figure the speed of sound at 23°C:

$$v = \left(\frac{\lambda kT}{m} \right)^{1/2}$$

$$v = \left(\frac{(1.40)(1.38 \times 10^{-23} \text{ kg·m}^2\text{s}^{-2}\text{K}^{-1})(296 \text{ K})}{4.80 \times 10^{-26} \text{ kg}} \right)^{1/2}$$

$$v \approx 345 \text{ m/s}$$

Tada! The speed of sound where you are is 345 meters per second. You can relate the time the signal took and the speed of sound to the distance this way:

Distance = speed × time

So how much time did it take for the sound to speed from your camera to your friend? Well, the camera recorded 4.00×10^{-2} seconds, but don't forget

that's the time for a round trip (the sound leaving the camera and then return-
ing after bouncing off your friend). So the time sound takes to reach your
friend is 4.00×10^{-2} seconds $\div 2 = 2.00 \times 10^{-2}$ seconds, which translates to

$$\text{Distance} = (345 \text{ m/s})(2.00 \times 10^{-2} \text{ s}) = 6.90 \text{ meters}$$

And that's that. Your friend was standing roughly 6.90 meters away from you
when you took the photo.

Faster: The speed of sound in liquids

Sound travels faster in liquids than it does in gases. That's because liquids
are less elastic than gases, meaning they "bend" less under the same applied
force. When you create a disturbance in a liquid, the force opposing that disturbance is greater in a liquid than in a gas, which means the liquid "snaps
back" into place quicker. The end result is that the disturbance is chased
through the liquid faster than it is in a gas.

So what's the expression for the speed of sound in liquids? That depends on
two main aspects of the liquid:

- ✔ **The resistance to deformation:** The speed of sound is a measure of how
  fast the medium "snaps back" into place after a disturbance, and the
  measure of that is closely tied to the medium's *bulk modulus* (the resistance of a substance to being deformed by pressure). In fact, it's tied to
  the adiabatic bulk modulus (*adiabatic* means no heat is exchanged with
  the environment), whose symbol is β_{ad}.

  The larger the adiabatic bulk modulus, the more resistance the liquid
  puts up against being deformed; so the higher the β_{ad}, the higher the
  speed of sound in that liquid.

- ✔ **The density:** The speed of sound in liquids is also tied to the liquid's
  density (ρ). The higher the density of the liquid, the harder it is to get
  the liquid to move. Thus, the speed of sound is lower in dense liquids.

Putting it all together, you get this for the speed of sound in a particular liquid:

$$v = \left(\frac{\beta_{ad}}{\rho} \right)^{1/2}$$

where β_{ad} is the adiabatic bulk modulus and ρ is the liquid's density.

Here's your chance to practice calculating the speed of sound in a liquid!
(Please contain your excitement.) Suppose you and your Physics II classmate
take a trip to the seaside, and you want to document the trip with a photo of
your friend. Unfortunately, she's scuba diving, so you need to go underwater
to take the photo. You set your camera to take underwater photos and take

the snapshot. Your friend sees you doing it and comes over, wanting to know how far apart the two of you were.

When you took the underwater photo of your pal, the camera said the sound signal came back in 4.00×10^{-3} seconds. Given that the adiabatic bulk modulus of water is 2.31×10^9 pascals and the density of water is 1,025 kilograms per cubic meter, how far away was your friend?

Calculate the speed of sound in water with this equation:

$$v = \left(\frac{\beta_{ad}}{\rho}\right)^{1/2}$$

$$v = \left(\frac{2.31 \times 10^9 \text{ Pa}}{1,025 \text{ kg/m}^3}\right)^{1/2}$$

$$v \approx 1,500 \text{ m/s}$$

So the speed of sound in water is about 1,500 meters per second. What does that info buy you? Well, you now know how long it took for the sound pulse from the camera to return to the camera, and you also know that distance = speed × time.

The camera recorded 4.00×10^{-3} seconds for a sound pulse to travel from the camera to your friend and back again, so the sound took $4.00 \times 10^{-3} \div 2 = 2.00 \times 10^{-3}$ seconds to reach your friend. Plugging in the speed of sound and the time the sound pulse took, you discover that the distance between you and your pal was

Distance = $(1,500 \text{ m/s})(2.00 \times 10^{-3} \text{ s}) = 3.00 \text{ m}$

Fastest: The speed of sound in solids

If the stiffer the medium is, the faster the speed of sound, then it shouldn't surprise you that sound travels fastest in solids, which are even less elastic than liquids.

So what's the expression for the speed of sound in solids? Here, you use a combination of

▶ *Young's modulus,* a measure of the stiffness of uniform materials

▶ The density of the solid

Here's how Young's modulus (Y) and density (ρ) relate to give you the speed of sound in a solid:

$$v = \left(\frac{Y}{\rho}\right)^{1/2}$$

This equation tells you that the higher Young's modulus is — in other words, the stiffer the medium — the faster the speed of sound. The greater the density of the material, the slower the speed of sound (because the material is slower to react to a disturbance).

Imagine that you're on an ocean cruise with your significant other. The two of you are standing on the deck, and because you both happen to be physicists, you naturally decide to measure the length of the deck, which is steel. Your significant other stands on the bow of the ship while you stand at the stern. Borrowing a handy fire axe from a fire control station on deck, you tap your end of the deck. Your significant other then reports that the sound took only 2.00×10^{-2} seconds to travel through the deck.

Given that Young's modulus for steel is $Y = 2.0 \times 10^{11}$ N/m^2 and that the density of steel is $\rho = 7,860$ kg/m^3, you can start determining the length of the deck by plugging numbers into the speed-of-sound expression for solids:

$$v = \left(\frac{Y}{\rho} \right)^{1/2}$$

$$v = \left(\frac{2.0 \times 10^{11} \text{ N/m}^2}{7,860 \text{ kg/m}^3} \right)^{1/2}$$

$$v \approx 5.0 \times 10^3 \text{ m/s}$$

So the speed of sound in steel is about 5.0×10^3 meters per second — that's 5 kilometers per second, or about 11,000 miles per hour.

You can find the length of the steel deck by multiplying the speed of sound by the time it took the sound to travel, which gives you

$$\text{Distance} = (5.0 \times 10^3 \text{ m/s})(2.00 \times 10^{-2} \text{ s}) = 100 \text{ m}$$

There you have it: The deck is about 100 meters long.

The solid sounds of the railroad tracks

As a kid, I was able to verify that sound travels faster in solids than through gases using railroad tracks that had those connector bars attached. By putting my ear to the track and watching a friend hit the tracks with a hammer some distance away, I could hear a definite clank-CLANK-clank coming through the tracks and then through the air. (**Note:** I don't recommend this as an experiment, especially if there are any trains lurking about.)

Analyzing Sound Wave Behavior

This section considers some of the weird and wonderful things that sound waves can do. You see them bouncing and bending and find out what happens when two sound waves meet. This leads you to discover a new kind of wave — the standing wave, which doesn't propagate; these are the kinds of waves that come from musical instruments. You see what happens when sources of sounds and listeners move. And finally, you break through the sound barrier to find out what happens when sound sources move faster than the speed of sound.

All the properties of sound that I discuss here are also properties of waves generally. So by understanding these aspects of sound behavior, you're actually getting a lot more in the bargain. For example, grasping sound waves can take you a long way toward an understanding of light and optics in the next few chapters. These wave properties go right to the heart of lots of the workings of the physical world.

Echoing back: Reflecting sound waves

Reflection occurs when a wave encounters a boundary. You're familiar with the reflection of sound waves in the form of an echo.

In the case of a sound wave in air, the condensation of the air in a high-pressure peak pushes against the air immediately next to it, which becomes condensed in turn, and so the wave spreads. However, if a high-pressure peak meets a solid surface such as a wall and tries to push against it, the wave doesn't find the wall so forgiving. The high pressure pushes the wall, but the wall pushes back with a force of resistance. The high-pressure peak now sits against the wall and pushes against the air behind it, which, being more forgiving than the wall, allows the high-pressure peak to propagate back the way it came — and the wave is reflected. Physicists say that the wall provides a *boundary condition* on the wave.

Sound is a *longitudinal wave,* in which the air molecules oscillate in the direction of motion of the wave. As the wave approaches the wall, the wall restricts the motion of the air molecules. The molecules right next to the wall can't oscillate at all. As the wave strikes, the molecules of the air near to the wall continue toward the wall until they effectively bounce off it — redirecting their motion in the opposite direction — thereby reflecting the wave. In these terms, the boundary condition on the wave is that there must be *zero oscillation* at the wall.

Seeing with sound

As you probably know, a bat "sees" not with reflected light but with reflected sound. When a bat hunts, it uses *echolocation;* it makes clicking sounds that bounce back from any unfortunate insects that may be flying by, and then it listens to the echo.

Bats had a head start, but as physicists came to understand the reflection of sound waves, people were able to use the same ideas for technologies like sonar and sonograms. These devices enabled people to see in a similar way, right down to the depths of the oceans and even inside the human body.

To illustrate this echoing process, I've graphed a pressure wave of sound as it reflects off a solid wall in Figure 7-5. For clarity, I haven't reflected a whole wave. I've sent just a part of a wave — a pulse — as a speaker would generate if its diaphragm moved out and then back just once. In this figure, x measures the distance from the wall and p is the pressure fluctuation.

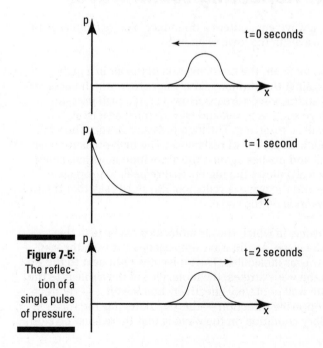

Figure 7-5: The reflection of a single pulse of pressure.

With a different type of boundary may come different boundary conditions. For example, the wave may fall on a soft wall, which is deformable. In this case, you may see some oscillation immediately next to the wall, and the wall may absorb some of the wave's energy as the molecules do work on the wall to move it. In that case, the reflected wave has a smaller amplitude, and the echo is quieter.

Sharing spaces: Sound wave interference

Two waves can occupy the same place at the same time. When this happens, they're said to *interfere*. The resulting oscillation is incredibly simple to work out: Just add the oscillation from one wave to the other. This idea is called the *principle of superposition*. So at any point, if one wave would cause a displacement of the medium of y_1 and the other wave would cause a displacement of y_2, then the actual displacement of the medium at that point is simply $y_1 + y_2$.

To see the principle of superposition in action, check out Figure 7-6. It shows the wave displacements over time for two separate waves. In the same graph, I show what would happen if those two waves were traveling through the medium at the same time.

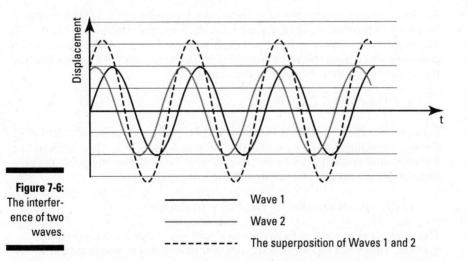

Figure 7-6:
The interference of two waves.

——————— Wave 1

——————— Wave 2

- - - - - - - - - The superposition of Waves 1 and 2

Adding amplitudes: Constructive and destructive interference

Interference can be constructive or destructive. With *constructive interference,* the amplitudes of two waves combine to make a wave of larger amplitude. With *destructive interference,* the amplitudes of the waves cancel each other out.

For instance, suppose you have a stereo with a pair of speakers like those in Figure 7-7. Now put on a CD that plays a pure tone — that is, each speaker makes the same sine-shaped sound wave (see the earlier section "Under pressure: Measuring the amplitude of sound waves" for details on why the graph takes this shape). This would be a very uninteresting piece of music to play, but it has some surprising effects.

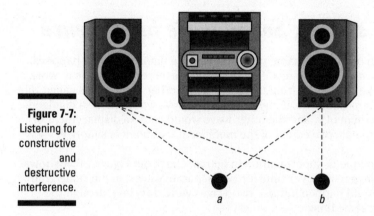

Figure 7-7:
Listening for constructive and destructive interference.

Say the speakers play a tone with frequency f, wavelength λ, and amplitude A. Now sit in a position equally distant from each speaker (point a in Figure 7-7). You receive two waves — one from each speaker. If you call the displacement you experience from the wave travelling from the left speaker y_1, you can say that the displacement is given by the sine wave

$$y_1 = A \sin(2\pi ft)$$

Because the other speaker is the same distance away, the displacement of the wave coming from it, y_2, is just the same as y_1, so $y_1 = y_2$. Now work out the wave that you experience, y_T, as you sit at point a. Use the principle of superposition and add the waves coming from each speaker:

$$y_T = y_1 + y_2 = A \sin(2\pi ft) + A \sin(2\pi ft) = 2A \sin(2\pi ft)$$

This is just a sine wave with twice the amplitude of the wave from each speaker. That's not too surprising — if you sit at point a, you just hear a louder sound than you would if you had just one speaker instead of two. The two speakers are combining to make a larger-amplitude wave — this is *constructive interference*.

Now suppose you move — sit just to one side (at point b in Figure 7-7) so that you're exactly half a wavelength closer to the right speaker than you are to the left one. This means that the wave from the right speaker reaches you half a period earlier than the wave from the left speaker — that is, it's shifted by $T/2$. So you can write y_2 as

$$y_2 = A \, \sin\!\left(2\pi f\!\left(t + \frac{T}{2}\right)\right)$$

$$= A \, \sin\!\left(2\pi ft + 2\pi f \frac{T}{2}\right)$$

$$= A \, \sin\!\left(2\pi ft + \pi\right)$$

$$= -A \, \sin\!\left(2\pi ft\right)$$

$$= -y_1$$

Now if you work out the combined wave from both speakers, you see

$$y_T = y_1 + y_2 = y_1 + (-y_1) = 0$$

You receive no sound wave at all — silence! The waves from each speaker are canceling each other at point *b* — this is *destructive interference*. You can read about constructive and destructive interference in light waves in Chapter 10.

Standing waves: Destructive interference at regular intervals

A *standing wave* is a kind of wave that doesn't travel — the peaks simply oscillate at the same place without propagating. This kind of wave occurs when a propagating wave is confined, such as on a piece of string or, as you see in this section, when sound is contained in a tube. Here, I show you how to construct a setup to contain the sound, and you see how sound reflects inside the tube and interferes to produce a standing wave.

The setup: Getting identical waves going in opposite directions

Suppose you take a long tube that's closed at one end and has a diaphragm at the other end (you stretch an elastic sheet over the end, for example). Place a speaker near the diaphragm. When you turn the speaker on, the sound waves cause the diaphragm to vibrate. The sound waves from the diaphragm travel down the tube (acting as the *incident wave*), reflect off the closed end, and travel back up the tube to the diaphragm again (the *reflected wave*).

But remember, the speaker does not produce a single pulse; instead, you have a sine wave of sound. So in this situation, you have two waves in the tube at the same time, one traveling away from the speaker and one traveling toward it. Imagine that the reflection is ideal so that both waves have the same amplitude, frequency, and wavelength; their only difference is that they're traveling in opposite directions.

Now look at the total wave. The wave from the diaphragm travels down the tube to the closed end, where the boundary condition is that there can be no displacement of the molecules. The displacement of the reflected wave, at the closed end, is always the opposite of the displacement of the incident wave. So when the waves interfere, there's no displacement (destructive interference) at all at the closed end, satisfying the boundary condition.

Because both waves are periodic, this destructive interference must happen at regular intervals along the tube. To see this, you move away from the wall by half a wavelength. Because both waves are sine waves, they both have the opposite displacement here that they had at the wall, so they're still both equal and opposite — you have destructive interference again. In this way, at every point along the tube that's a whole-number of half-wavelengths from the closed end, there's destructive interference, so the molecules do not oscillate at all in these places. The total wave in this tube must be different from the sine wave you usually see for sound — you get the full picture of this strange new wave next.

Graphing a standing wave

Figure 7-8 shows a graph of incident and reflected sound waves at several different times. They're two identical waves, with the only difference being that they're travelling in opposite directions. The incident and reflected waves have amplitude A, and the horizontal axis measures distance from the wall.

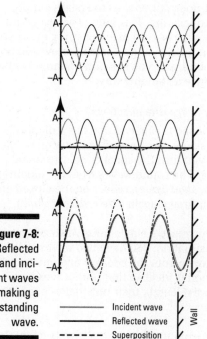

Figure 7-8:
Reflected and incident waves making a standing wave.

Incident wave
Reflected wave
Superposition

Wall

The figure also shows the interference between these two waves, which is just the sum of the two graphs (according to the principle of superposition). You can see this wave at three different times. Notice that the wave doesn't travel anywhere; this is a *standing wave,* and it just stays where it is. It oscillates, but the peaks and troughs do not propagate; they just move up and down.

The part of the wave that crosses the axis (where there's no displacement) is called the *node.* Because the wave doesn't propagate, its nodes don't move. These nodes are just points of destructive interference.

Between the nodes are points in the wave that oscillate with the greatest amplitude — these are the *antinodes.* The amplitude of the standing wave at these antinodes is just equal to twice the amplitude of the incident and reflected waves. So you have a picture of the standing wave as a nonpropagating oscillation, which has points of maximum and zero oscillation at intervals of $\lambda/2$ along its length.

Harmonics: Putting the standing wave in normal mode

When the oscillation of a diaphragm coincides with the antinode (greatest amplitude) of a standing wave, the standing wave is in a *normal mode.* Normal modes occur wherever there are standing waves, such as the ones on a vibrating string or in the pipes of an organ.

Suppose you have the closed-tube setup I describe earlier in "The setup: Getting identical waves going in opposite directions." You set the speaker to make a pure tone with a wavelength that makes a standing wave in the tube, which has an antinode at the diaphragm. The oscillation of the diaphragm coincides with the antinode of the standing wave, so it's in a normal mode.

Standing waves are in a normal mode when the speaker makes a sound with wavelength λ_n, which is given by

$$\lambda_n = \frac{4L}{n} \qquad (n = 1, 3, 5, \ldots)$$

where L is the length of the tube and n is a whole number that labels the various normal modes.

The frequencies of normal modes of vibration are called the *harmonics.* The first harmonic ($n = 1$) is called the *fundamental frequency.* Musicians often call the frequencies of the higher frequency modes *overtones.*

Note that n must be an odd number because there are an odd number of quarter-wavelengths from the barrier (the closed end of the tube) to an antinode. As n increases, so does the number of nodes and antinodes in your normal mode. So when your speaker makes a sound with wavelength $4L$, there's only one antinode, which is the one at the diaphragm.

Figure 7-9 shows some normal modes of your tube — the two lines show the two positions of maximum displacement of the normal mode. Here, the horizontal axis measures distance from the diaphragm, and you can see the position of the closed end of the tube on the right side of the graph.

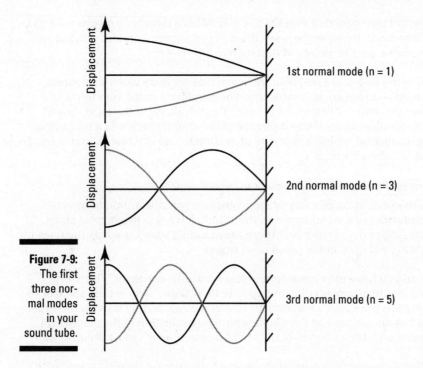

Figure 7-9: The first three normal modes in your sound tube.

Now try a concrete example — what are the first few notes that your tube likes to play? Suppose your tube is 0.983 meters long. Then the wavelength of its normal modes are

$$\lambda_n = \frac{4}{n}(0.983 \text{ m})$$

If the speed of sound in your tube is 343 meters per second, then the frequencies of these modes, f_n, are

$$f_n = \frac{v}{\lambda_n} \qquad (n = 1, 3, 5, \dots)$$

$$= \frac{343 \text{ m/s}}{\frac{4}{n}(0.983 \text{ m})}$$

$$\approx n(87.2 \text{ Hz})$$

This means that the frequency of the lowest normal mode is 87.2 hertz.

The next normal mode, when $n = 3$, has a frequency of 262 hertz, which is about middle C on a piano. Where would you have to place your ear in the tube in order to hear silence when you play middle C on your speaker? You'd just have to listen at a node, which happens every half-wavelength from the closed end of the tube. When $n = 3$, your wavelength, λ_3, is given by

$$\lambda_3 = \frac{4}{3}\left(0.983 \text{ m}\right) \approx 1.31 \text{ m}$$

So you'd have to place your ear half this distance from the closed end of the tube — that is, 0.655 meters. For this normal mode, there are no other nodes in the tube (except, of course, the one at the closed end of the tube), so this is the only place you'd get silence.

It turns out that *any* possible vibration of sound in the closed-tube-and-diaphragm setup is simply an interference of normal modes! So even the craziest, most complicated, erratic vibration can be boiled down to a matter of how much of each normal mode you have. This understanding comes from some extremely powerful ideas in mathematics that have pervaded physics. For example, in quantum mechanics, particles (like the electron) are allowed to be only in certain particular states. These states are like the normal modes of your tube. Like your tube, the particles can be in a state that's an interference of these normal modes — but when you actually measure the state of the electron, say, you can only ever see it in one of the normal modes! This is just a hint of some of the quantum weirdness that lies ahead for you in physics.

Reaching resonance frequency: The highest amplitude

You can drive things at a frequency that maximizes the amplitude of vibration. For instance, consider the sound vibration in the speaker-and-tube setup, which is driven by the speaker. As you increase the frequency of the sound wave from the speaker, you find that the amplitude of the sound vibration peaks whenever the speaker drives at one of the harmonics — that is, one of the frequencies of the normal modes. So the tube has an infinite number of resonance frequencies, given by f_n, where n is an odd number.

It's testament to the power of the ideas in physics that resonance is also a feature of the RLC electrical circuit that you look at in Chapter 5. You see that in the RLC circuit, there's likewise a natural frequency at which the current in the circuit oscillates with the greatest amplitude.

Getting beats from waves of slightly different frequencies

Anyone who has tuned a guitar has heard the effect of simultaneously playing two very slightly different notes — the sound seems to oscillate in loudness. These oscillations are called *beats*.

Figure 7-10a shows a graph of two waves of slightly different frequencies, and Figure 7-10b shows their sum, which is the wave of interference between the two. You can see that the interfering wave oscillates with a frequency similar to the frequency of the original two waves, but the amplitude increases and decreases with another frequency, the beat frequency. The *beat frequency* is simply the difference between the frequencies of the original waves.

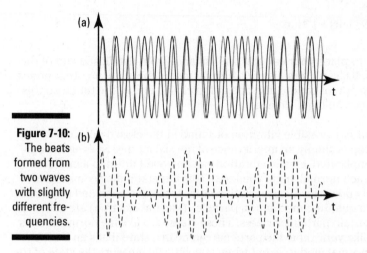

Figure 7-10: The beats formed from two waves with slightly different frequencies.

Bending rules: Sound wave diffraction

You can hear a police car approaching with its siren wailing even if it's around a corner, hidden by tall buildings. And you can talk to a person in another room through an open door, even if you can't see that person through the doorway. Sound waves travel in straight lines, but when they hit a boundary like the edge of a wall or a lamppost, the sound waves bend around it — this behavior is *diffraction*.

Diffraction happens in all waves, including sound. Figure 7-11 shows a sound wave approaching two gaps in a wall — the lines represent the position of the wave peaks. One gap is much wider than the wavelength of the sound, and the other is similar in size to the wavelength. You see that as the sound wave travels through the wider gap, it mostly goes straight through, with some bending at each edge. But when the sound goes through the gap that's of similiar width to the wavelength of the sound, the wave spreads over a wider angle. This bending of the wave, to where it wouldn't go if it traveled in a straight line, is diffraction.

The wider angle you get when a wave spreads by diffraction explains why you can hear around corners but you can't see around them. Light has a much shorter wavelength than sound, so if you were to shine a light through the gaps in Figure 7-11, both gaps would be much wider than the wavelength of the light; therefore, you'd see hardly any spreading of the light wave — not enough for you to notice, anyway.

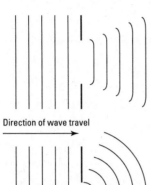

Direction of wave travel

Figure 7-11:
Diffraction
of a sound
wave
through a
gap in a
wall.

Diffraction is really just a manifestation of interference (as I show you in Chapter 10, on light). People use two different terms for essentially the same thing, but the difference is that interference is usually understood to mean the interaction of just a few waves, whereas diffraction is the interference of a very great number of waves.

Coming and going with the Doppler effect

The *Doppler effect,* named after Christian Doppler, says that a sound wave's frequency changes if the source of the sound is moving (or you're moving toward or away from the source). If you and the source of the sound are getting closer, you hear the sound at a higher pitch. And if you and the source are getting farther apart, you hear the sound at a lower pitch.

For instance, consider a police car with its wailing siren. Because you're law-abiding, it passes right by you. What do you hear? You're familiar with the high-pitched *ne-naw-ne-naw* as the car travels toward you, turning into a low-pitched version of the same sound after it has passed and is traveling away. You can understand this effect using the picture of sound as a wave.

Moving toward the source of the sound

First consider what happens when the source of sound is stationary but you're moving toward it. You can see this situation in Figure 7-12a. The source produces a wave with wavelength λ_s and frequency f_s, and this wave travels with the speed of sound v. You walk toward the source with speed v_a.

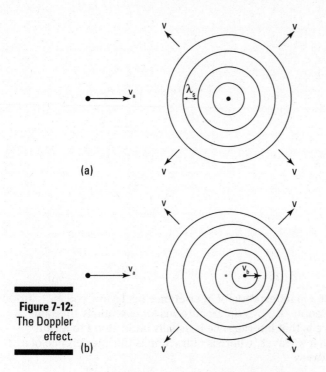

(a)

(b)

Figure 7-12:
The Doppler
effect.

If you were to remain stationary, your ear would experience a sound wave of frequency f_s, which is the frequency of the source. But if you walk toward the source, then your ear experiences a sound wave with a higher frequency. You're walking into the wave, so each wave peak has to travel a slightly smaller distance to reach you than it would if you were to remain still.

As you move toward the source of the sound, the speed of the wave as it appears to you is $v + v_a$. So when a wave peak is at your ear, the time for the next wave peak to reach you is $\lambda_s/(v + v_a)$ seconds. Therefore, the frequency you hear, f_a, is given by

$$f_a = \frac{v + v_a}{\lambda_s}$$

Because $v = \lambda_s f_s$, you can write this as

$$f_a = \frac{v + v_a}{v} f_s = \left(1 + \frac{v_a}{v}\right) f_s$$

So you see that the frequency you hear is a factor of $(1 + v_a/v)$ greater than the frequency from the source.

Having the source of the sound move

When the source of a sound moves, the speed of the sound waves remains the same, because the speed of sound is determined only by the air — it has nothing to do with the source. So if the source moves away from you with a speed v_b, as in Figure 7-12b, v is still the same. But what changes is the wavelength of the sound waves.

To see why the wavelength changes, think about how the waves propagate. Say the source emits a wave peak, which propagates behind the source. In the time before the next peak, the source moves a distance $v_b T_s$, where T_s is the period of the source waves ($T_s = 1/f_s$). So the wavelength behind the source is

$$\lambda = \frac{v}{f_s} + \frac{v_b}{f_s} = \frac{v + v_b}{f_s}$$

The wavelength of the wave behind the source is now enlarged. Now plug this new wavelength into the equation relating the frequency that you hear to the frequency of the source (from the preceding section). Replace λ_s with the new enlarged wavelength to find

$$f_a = \frac{v + v_a}{\lambda} = \frac{v + v_a}{v + v_b} f_s$$

This is the frequency you hear when you travel toward a moving source of sound, which is in front of you. The source is moving in a particular direction with speed v_b, and you're behind it traveling in the same direction with speed v_a.

If you're in front of the source, then you're in the region where the waves from the source have a shorter wavelength, and you're walking away from the approaching peaks. The frequency you hear is given by

$$f_a = \frac{v - v_a}{\lambda} = \frac{v - v_a}{v - v_b} f_s$$

Doing the math on the Doppler effect

Now put some numbers into a police siren example. Suppose that the police car passes very close to you. It's initially traveling pretty much straight toward you, and after it passes, it travels pretty much straight away from you.

A police siren makes a sound that has a frequency of about 320 hertz. If the car has its siren on, it must be in a hurry, so say it's going at about 70 miles per hour (31.29 meters per second). Also, you're walking along the sidewalk at a speed of 1.5 meters per second. The actual frequency of the sound you hear is

$$f_a = \left(\frac{343 \text{ m/s} - 1.5 \text{ m/s}}{343 \text{ m/s} - 31.29 \text{ m/s}} \right) 320 \text{ Hz}$$

$$= \left(\frac{341.5}{311.71} \right) 320 \text{ Hz}$$

$$\approx (1.0956) 320 \text{ Hz}$$

$$\approx 351 \text{ Hz}$$

This is about a 10 percent increase on the actual frequency of the siren. Now work out the frequency you hear when the police car has passed and is traveling away:

$$f_a = \left(\frac{343 \text{ m/s} + 1.5 \text{ m/s}}{343 \text{ m/s} + 31.29 \text{ m/s}} \right) 320 \text{ Hz}$$

$$= \left(\frac{344.5}{374.29} \right) 320 \text{ Hz}$$

$$\approx (0.9204) 320 \text{ Hz}$$

$$\approx 295 \text{ Hz}$$

This is about an 8 percent decrease on the actual frequency of the siren.

If you have a piano handy, you can play these sounds. The original police siren is roughly the same tone as the *E*, which is two whole notes above middle *C*. Then the sound that you hear as the car travels toward you is about the same as playing one note higher, and the sound when the car has passed is about the same as one note lower.

Breaking the sound barrier: Shock waves

Sound moves pretty quickly in air, but some things move faster than sound. When Concorde (the French and British supersonic passenger jet) was flying before its retirement in 2003, you could travel across the Atlantic Ocean at about twice the speed of sound. Meteors entering the Earth's atmosphere travel through the air much faster than this. Objects breaking the sound barrier produce a *sonic boom,* a loud sound that people can hear from the ground. In this section, I discuss what happens when something breaks the sound barrier.

Producing shock waves

Because of the Doppler effect, the wavelength of the sound produced by a moving source is stretched behind it and shortened in front (see the earlier section "Coming and going with the Doppler effect" for details). When an airplane (or another moving object) approaches the speed of sound, it has to work extra hard to compress the air in front of it as it bunches up all the wave peaks; this extra work gave rise to the term *sound barrier*.

Figure 7-13 shows the wave peaks of a source of sound moving faster than the speed of sound (that is, the source is in *supersonic motion*). The wave peaks spread uniformly away from where the source was when it emitted them. At any one time, this makes a series of circles whose centers are evenly spaced along the path of the source (assuming it's moving at a constant speed), and the radii of the earliest circles are uniformly greater than the most recent ones. The edge of these circles forms a line of constructively interfering waves along their outer edge, which is called a *shock wave*.

The concentration of constructively interfering sound waves along the shock wave makes a very loud sound. Any listener who happens to be at a point on the shock wave hears this sound — the *sonic boom*.

Note that as the plane travels through the air, the shock wave is actually a cone; the figure shows only a cross section. As the plane travels faster than sound and produces a shock wave, the air around the tip of the plane is at a higher pressure than the surrounding air. The speed of sound can vary as the pressure of the air varies, which means that the pressure variation around the tip of the plane causes the shape of the shock wave to curve slightly in this region instead of the straight lines you see in Figure 7-13.

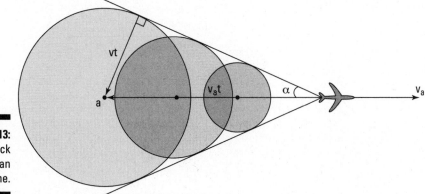

Figure 7-13:
The shock wave of an airplane.

Finding the angle of a shock wave

With basic trigonometry, you can easily work out a good approximation of the angle a shock wave makes from the direction of travel. Look at the right triangle in Figure 7-13. Since the source (an airplane) emitted a sound wave at point a, the source has been traveling for a time t at speed v_a. So the length of the hypotenuse of this triangle is $v_a t$, and the length of the opposite side of the triangle is just the distance that the sound wave has traveled, vt. The sine of the angle of the shock wave, α, is then just the ratio

$$\sin(\alpha) = \frac{vt}{v_a t} = \frac{v}{v_a}$$

You may hear that a jet travels at such-and-such a Mach number, such as Mach 3.3 (the SR 71 Blackbird) or Mach 9.6 (NASA's X-43A). The *Mach number* is just the speed of the jet compared to the speed of sound, v_a/v.

Check out an example — what would be the angle of the shock wave that Concorde would've made as it traveled across the Atlantic at twice the speed of sound? This means that the Mach number is 2.0; hence,

$$\sin(\alpha) = \frac{v}{v_a} = \frac{1.0}{2.0} = 0.50$$

If you take the inverse sine of this, you find the angle is 30°.

Chapter 8

Seeing the Light: When Electricity and Magnetism Combine

Cracking the secret of light was a major advance for both scientists and the general population. Now physicists know what creates light waves. They can even predict how fast light waves go, how much energy they transfer from Point A to Point B, and more. Consider this chapter to be your guided tour of the nature of light. I'm your friendly guide (minus the name badge), and I start things off by covering what light really is. You then dive into topics such as the electromagnetic spectrum, light intensity, and more.

Let There Be Light! Generating and Receiving Electromagnetic Waves

The big name in the discovery of how light works is James Clerk Maxwell. He's the lucky physicist who first figured out that *light* is nothing more than alternating electric and magnetic fields that regenerate each other as light travels, allowing it to keep going forever.

In this section, I explain how electricity and magnetism combine to create electromagnetic waves. I also note how radio receivers work by catching either the electric or the magnetic field of those waves.

Creating an alternating electric field

The process of generating an alternating electric field (known as an *E field*) starts with an oscillating charge. To create an oscillating charge, you can connect an alternating voltage source to the top and bottom of a wire. Figure 8-1 shows this setup at four consecutive times. The alternating voltage source causes the electrons in the wire to race up and down its length, creating an alternating electric field in the wire.

Electric fields propagate through space, so as the electric field you created goes up and down the wire, that same electric field moves out into space as well. Because the electric field in the wire is constantly changing directions as the voltage source alternates, you get an alternating electric field in the wire, which leads to an alternating electric field propagating through space, as you see in Figure 8-1.

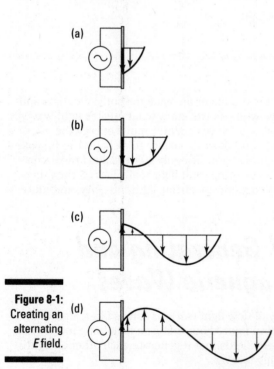

Figure 8-1: Creating an alternating *E* field.

At first, the electric field starts off small (see Figure 8-1a). Consequently, the electric field that leaves the wire and propagates through space is also small. In time, however, the electric field in the wire becomes larger (Figure 8-1b), and the propagated electric field does the same.

Then, as the voltage source alternates, the electric field begins to switch directions in the wire. The propagated electric field follows, growing smaller even though it's still pointing in the same direction. At some later time, the voltage across the wire changes polarity completely (the *polarity* of the potential difference between two points just describes which point is of higher potential and which is lower) — and the electric field in the wire changes polarity, too, as you can see in Figure 8-1c. Not surprisingly, the direction of the electric field that's propagated into space also changes.

As you can see in Figure 8-1d, as the oscillating charge completes its alternating cycle, the wave in the electric field completes its cycle, too.

Note how the oscillating electric field in Figure 8-1 always points in a direction that's perpendicular to the direction of propagation. The wave propagates to the right, with the electric field always in the vertical orientation — that is, it oscillates up and down. When the electric field oscillates with a constant orientation as the wave propagates, the wave is *linearly polarized*. So you can say that the wave propagating in Figure 8-1 is linearly polarized with the electric field in the vertical orientation. (Why not bring it up at your next party?)

Getting an alternating magnetic field to match

How exactly do you pair an alternating electric *(E)* field (see the preceding section) with the magnetic *(B)* field that's supposed to be the other half of a light wave? Are you supposed to have a spinning bar magnet or something?

Actually, creating the matching *B* field is easier than it seems. In fact, if you have a straight wire where the voltage (and hence the current) alternates up and down, you've already done it, because current in wires creates a magnetic field.

Here's how applying an alternating voltage across a wire creates your matching *E* and *B* fields:

✔ When you have *voltage* alternating up and down a wire, you create an oscillating *E* field.

✔ That *E* field causes *current* to race up and down in the wire, and the current generates an alternating *B* field.

Notice the direction of the created *B* field in Figure 8-2. The created *E* field is parallel to the wire (following the electrons as they race madly up and down the wire), but the *B* field is perpendicular to the *E* field. That's because the *B* field is created perpendicular to the wire.

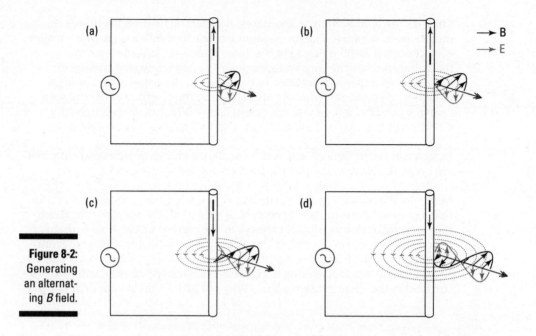

Figure 8-2:
Generating
an alternat-
ing *B* field.

Putting all this together means that an alternating voltage source applied
across a wire creates an alternating *E* field and an alternating *B* field, both
of which propagate away from the wire, as Figure 8-3a shows. Note that the
E and *B* fields are perpendicular to each other. And both of them, the *E* field
and the *B* field, are perpendicular to the direction of propagation — you have
all three dimensions covered.

When you know the directions of the *E* and *B* fields, just follow one of the
right-hand rules to find the direction of propagation:

  ✓ If you put the fingers of your right hand in the direction of the *E* field and
    then bend them toward the *B* field using the shortest possible arc, then
    your thumb will point in the direction of propagation.

  ✓ Hold out your palm, pointing your fingers in the direction of the electric
    field and your thumb in the direction of the magnetic field. Then the
    direction of propagation of the wave is the direction your palm is facing.
    Figure 8-3b shows this version.

Electromagnetic waves are just propagating fluctuations of the electric and
magnetic fields. Electromagnetic waves of the lowest frequencies — like the
ones from a wire connected to an alternating voltage source — are *radio
waves.* A higher range of frequency of electromagnetic waves is even more

familiar: light. That's right — light and radio are essentially the same thing; the only difference is that your eyes are sensitive to the frequencies of visible light waves.

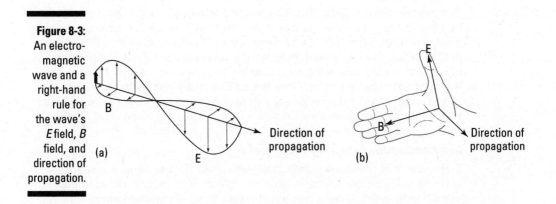

Figure 8-3:
An electro-
magnetic
wave and a
right-hand
rule for
the wave's
E field, *B*
field, and
direction of
propagation.

The wire in my example, believe it or not, is actually an *antenna*. Perhaps you've seen radio towers that soar up to great heights. At their core, they rely on a single wire with an alternating voltage placed across that wire from top to bottom. The wire, by having charges race up and down its length, creates radio waves.

Can you generate visible light with a wire in the same way that you can generate radio waves? Probably not. No alternating voltage source in the world oscillates fast enough to approach the frequencies of visible light. For the scoop on visible light and other parts of the electromagnetic spectrum, check out the later section "Looking at Rainbows: Understanding the Electromagnetic Spectrum."

Receiving radio waves

Creating radio waves (see the preceding sections) is only half the story; you still need a way of receiving them. That's where receiving antennas come in.

As I show you in Figure 8-3, the electric and magnetic fields of a radio wave are perpendicular to each other — there, the *E* field moves vertically and the *B* field moves horizontally. Vertical antennas and loop antennas are the two primary ways of receiving radio waves, and they correspond to the *E* and *B* field parts of radio waves, respectively.

Vertical antennas: Catching the E field

To detect a vertically moving electric field from a sending antenna, you simply use a vertical receiving antenna, which is really just a long wire.

The electric field (E field) from the sending antenna is in the vertical plane, just like the antenna itself, because the E field follows the movement of the electrons in the wire (see the earlier section "Creating an alternating electric field"). When you use a vertical receiving antenna, the E-field component of the radio wave makes the electrons in the receiving antenna race up and down. When the antenna receives the E field, a tiny voltage appears from the top to the bottom of the receiving antenna. Your radio can then amplify that voltage until it becomes a signal that lets you make out words and music.

Loop antennas: Catching the B field

To receive a horizontally moving magnetic field (B field) from a sending antenna, you can use a wire loop or coil. (***Note:*** Receiving radio antennas use a combination of both loops and coils.) First, set up the loop or coil in the vertical plane to maximize the magnetic flux through it (I cover magnetic flux in detail in Chapter 5). If that sounds counterintuitive to you, consider this: The magnetic field you're trying to detect is in the horizontal plane, so setting up a loop or coil of wire vertically allows you to make as much of that magnetic field as possible go through your antenna.

So the rapidly oscillating B field is oscillating in your loop or coil of wire. That's great, but how do you actually measure that B field? A changing magnetic flux in a loop or coil of wire induces a current in that loop or coil in a way that counteracts the applied magnetic field from the radio station. Your radio is able to measure that tiny current and decipher it, just as other radios can decipher the tiny voltages created by the radio station's electric field.

Making radio waves a hit

Physicist Heinrich Hertz was the one who first generated and received radio waves in his laboratory in 1886. This was a breakthrough for physics, but he wasn't sure how to put these waves into use.

Guglielmo Marconi, an Italian physicist, was one of many people who set out to use these new waves to communicate over great distances almost instantaneously. He patented a version of the telegraph that marked one of the first practical advances in "wireless" communication.

Early in radio's development, radio waves were detected over distances of about a mile. But physicists soon noted that the more charge racing up and down the antenna, the greater the amplitude of the wave and therefore the greater the distance from which it could be received. As transmitters and receivers advanced in technology, the distance increased to hundreds and then thousands of miles.

With a loop antenna, your radio decodes the current that flows through the loop due to the fluctuating magnetic flux. With a straight antenna, your radio decodes the tiny voltages that appear across the wire from the electric field component of the wave.

Looking at Rainbows: Understanding the Electromagnetic Spectrum

Electromagnetic waves have the same general properties shared by all waves — wavelength, frequency, and speed (see Chapter 6 for details). In this section, you see how those properties apply to light waves. You also find out how the continuous range of frequencies is divided up into different wave types within the electromagnetic spectrum.

Perusing the electromagnetic spectrum

Even though all electromagnetic waves are essentially the same — differing only in frequency — they differ in how they interact with matter. For example, the waves with a particular range of frequencies are visible as light, whereas other waves at a higher frequency are invisible but can give you a nasty sunburn. You see this variation because matter is made up of charged particles (electrons and protons) in various configurations, and the way that these particles interact with electromagnetic waves depends on the details of this configuration.

Different frequencies of electromagnetic waves correspond to different parts of the *electromagnetic spectrum* — that is, the range of all electromagnetic waves, arranged in increasing frequency. Most divisions of the spectrum are made according to how the different parts of the spectrum interact with matter, but the division is sometimes based on how the wave is produced or used.

People sometimes debate which wavelengths go in which category, but Figure 8-4 can give you approximate ranges of the main divisions of the electromagnetic spectrum, with labels for the names of the electromagnetic waves within them.

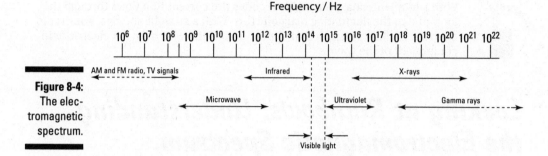

Figure 8-4:
The elec-
tromagnetic
spectrum.

Starting at the lower frequencies, here are the types of electromagnetic waves in order:

- ✔ **Radio waves:** As you can see in Figure 8-4, radio waves include the familiar AM and FM bands. The radio band's AM frequencies are in the 10^6 hertz (Hz) region, and the FM frequencies are in the 10^8 Hz region. Radio waves have long wavelengths and are generally produced with antennas (see the earlier section "Let There Be Light! Generating and Receiving Electromagnetic Waves" for details).

- ✔ **Microwaves:** When the frequency of radio waves increases to the point where the wavelength is about the same size of the electrical circuits used to make them, the wave can have a feedback effect on the circuit. The methods of generating waves of this frequency have to take this into account, so these waves have a special name: *microwaves.* Some liquids consist of molecules that absorb microwaves and become heated, which microwave ovens take advantage of.

- ✔ **Infrared light:** This kind of light is invisible to the naked eye. Humans have to wear night-vision goggles to pick up this part of the spectrum.

- ✔ **Visible light:** The light you *can* see is actually a very narrow band of the spectrum that exists solely in the 4.0×10^{14} Hz to 7.9×10^{14} Hz region (this is one of the few frequency ranges pretty much everyone agrees on). The lowest-frequency end of this part of the spectrum corresponds to the red end of the rainbow, and the highest-frequency end corresponds to the violet part of the rainbow. The rest of the rainbow is distributed within this range.

Why is visible light restricted to such a narrow range? One answer is that much of the rest of the light spectrum is absorbed by water and water vapor — both of which are plentiful on Earth. Infrared light, for example, is absorbed by water vapor, making it an unfavorable option to rely on for your vision.

- ✔ **Ultraviolet light:** Higher in the spectrum still, you have ultraviolet light, where the frequency is higher and the wavelength is shorter. This is the region of the so-called black lights that make phosphorescent paints glow. These are also the waves responsible for sunburn.

✔ **X-rays:** This part of the light spectrum travels easily through the human body, which is why X-rays play such a starring role in medicine to check for broken bones.

✔ **Gamma rays:** These high-energy rays are created by high-power transitions in the atomic nucleus (as opposed to other kinds of electromagnetic waves, which mostly come from transitions in the electron structure of an atom).

Relating the frequency and wavelength of light

Because light is made up of electromagnetic waves, it must obey the general wave equations (see Chapter 6). In particular, you can relate the frequency (*f*) of a wave to its wavelength (λ) to find its speed (*v*) like this:

$$v = f\lambda$$

In a vacuum, light travels at the speed *c*, which is about equal to 3.0×10^8 m/s (I explain where this number comes from in the next section). So for a vacuum (or air), you can say the following:

$$c = f\lambda$$

So using this formula, what's the wavelength of red light if its frequency is 4.0×10^{14} Hz? And at the other end of the visible spectrum, what's the wavelength of violet light (whose frequency is 7.9×10^{14} Hz)? You know that $c = f\lambda$, so the wavelength formula is

$$\lambda = \frac{c}{f}$$

Plugging in the numbers and doing the math for the red-light question gives you this:

$$\lambda = \frac{3.0 \times 10^8 \text{ m/s}}{4.0 \times 10^{14} \text{ Hz}} = 7.5 \times 10^{-7} \text{ m}$$

Now take a look at the calculations for violet light, where the frequency is 7.9×10^{14} Hz:

$$\lambda = \frac{3.0 \times 10^8 \text{ m/s}}{7.9 \times 10^{14} \text{ Hz}} \approx 3.8 \times 10^{-7} \text{ m}$$

The *nanometer* (abbreviated *nm*), or 10^{-9} m, is often used for wavelengths in the visible region. So you can say the two wavelengths are 750 nanometers and 380 nanometers. What do these numbers actually mean? Turns out that's up to your eye.

Red light is the longest wavelength your eye can perceive, and 750 nanometers is the longest of the red wavelengths most eyes can see. Violet is the shortest wavelength of light you can see, and 380 nanometers is the shortest of the violet wavelengths your eye can normally pick up. So in between 380 and 750 nanometers — a very short range — lie all the glorious colors of the light spectrum that are visible to the human eye.

See Ya Later, Alligator: Finding the Top Speed of Light

Light is fast — nothing can travel faster, *Star Trek* and *Star Wars* gadgetry included, unfortunately. The speed of light in a vacuum is approximately 3.0×10^8 meters per second, or 3.0×10^{10} centimeters per second, or about 186,000 miles per second. (If you're a stickler for accuracy, try the value 299,792,458 meters per second.)

The distance around the world is about 40,000 kilometers, or 4.0×10^7 meters, so at the speed of light, you could make 7.5 trips around the world in 1 second (3.0×10^8 m/s $\div$ 4.0×10^7 m = 7.5 trips/second). You could even go to the moon in that amount of time. So although light is fast, it's not infinite.

A not-so-illuminating light experiment

There was a time, of course, when people had no idea how fast light was. Many experiments were tried, and many failed (utterly). Case in point: In a touching show of confidence, two scientists synchronized their pocket watches to within a second and then trooped to opposite ends of a mile-long field. At just the agreed-upon moment, the first scientist opened a lantern. The problem was that from the second scientist's point of view, as soon as his watch showed the correct time, the beam of light from the first scientist was already shining full force. Neither scientist could believe anything could be so fast; each one thought his watch must've been off!

Of course, given the speed of human reflexes, the two scientists could've been standing 100,000 miles apart, and the beam of light would've arrived within less than a second — that is, less than the accuracy of the watches and the scientists' ability to open their respective lanterns.

In this section, you discover how physicists figured out how fast light travels in a vacuum. Of course, as with other waves, the speed of light depends on the medium it's traveling through, if any. I touch on light as it travels through materials such as diamond and glass in Chapter 9.

Checking out the first speed-of-light experiment that actually worked

Many people attempted to measure the speed of light, often relying on astronomical phenomena. Armand Fizeau and Léon Foucault were the first to make Earth-bound measurements of the speed of light. Foucault's method used a rotating mirror to improve upon the space-based estimates.

Albert Michelson, an American who adapted and improved upon Foucault's method, measured the speed of light in 1926 — and dramatically increased the accuracy of the measurements.

Setting up the experiment

Michelson's apparatus was pretty clever; it involved bouncing light off a mirror 35 kilometers away. However, because light makes the 70-kilometer round trip in about a ten-thousandth of a second, Michelson needed to do more than just bounce light off a mirror some distance away.

His solution made him famous, and you can see a depiction of it in Figure 8-5. To accurately capture the speed of light, Michelson determined that in addition to bouncing off a mirror 35 kilometers away, light had to hit a rotating, eight-sided mirror just right. Specifically, the light needed to bounce off one side of the eight-sided mirror, make a round trip of 70 kilometers, and then hit another part of the eight-sided mirror just right to get into the detector. If the mirror rotated too much or too little, the side that the light signal was meant to bounce off of into the detector just wouldn't be there (in other words, it wouldn't have reached its proper position yet). Because Michelson could regulate how fast the mirror rotated, which was pretty darn fast, he was able to make the window for light to hit the eight-sided mirror very small. Pretty clever, eh?

In the 1926 round of experiments, Michelson determined the speed of light to be 299,796 kilometers per second, plus or minus 4 kilometers per second. (However, 3.0×10^8 meters per second is sufficiently accurate for the calculations in this book.)

Finding the mirror speed

Try calculating how quickly Michelson's mirror must've been rotating to capture the speed of light. Say you're working with the setup in Figure 8-5 and want to reflect a light beam off a mirror 35 kilometers away (that's a round trip of 70 kilometers). The key to calculating the speed of the mirror's rotations is to realize that the shortest time the experiment can measure is the amount of time it takes for the eight-sided mirror to make one-eighth of a revolution. That's the shortest time the light can take to bounce off the far mirror, return, and still enter the detector.

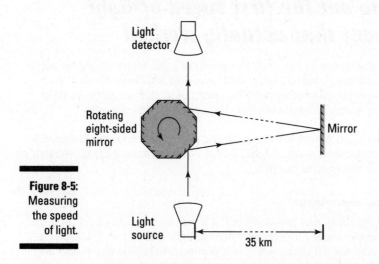

Figure 8-5:
Measuring
the speed
of light.

So in the time it takes light to go 70 kilometers, your eight-sided mirror makes one-eighth of a revolution. How fast does the eight-sided mirror turn? First, figure out the amount of time light needs to go 70 kilometers. You already know that

$$\text{Speed} = \frac{\text{distance}}{\text{time}}$$

So this equation must also be true:

$$\text{Time} = \frac{\text{distance}}{\text{speed}}$$

To travel 35 kilometers to the mirror and 35 kilometers back at the speed of light, you need this much time:

$$\text{Time} = \frac{2\left(3.5 \times 10^4 \text{ m}\right)}{3.0 \times 10^8 \text{ m/s}} \approx 2.3 \times 10^{-4} \text{ s}$$

That means your eight-sided mirror must make one-eighth of a turn in 2.3×10^{-4} seconds, giving it an angular speed of

$$\omega = \frac{\frac{1}{8}\text{revolution}}{2.3\times10^{-4}\text{ s}} \approx 540 \text{ revolutions/s}$$

So your eight-sided mirror needs to be turning at 540 revolutions per second in order to accurately measure the speed of light.

Calculating the speed of light theoretically

As James Clark Maxwell discovered, the astonishing fact is that absolutely every property of electric and magnetic fields — every aspect of their behavior — is contained in just four equations. Most of the math goes beyond trig, so you can skip the actual equations for now, but here's a preview of what they cover (see Chapters 4 and 5 for more on electric and magnetic fields):

- ✔ Faraday's law describes the electric field that comes from a changing magnetic field.

- ✔ Ampère's law describes the magnetic field that results from a current and a changing electric field.

- ✔ A third equation simply expresses the fact that there are no magnetic monopoles, so magnetic field lines are all loops.

- ✔ Gauss's law describes the flux of the electric field in terms of the electric charge. (For uniform fields, the *electric flux* of a field through an area is simply the size of the area multiplied by the size of the component of the field that's perpendicular to the area.)

Maxwell summarized and organized all the laws of electricity and magnetism because he was trying to solve a great puzzle. Before Maxwell came along, electric charge was thought to be divided — there was one form of charge for static electric fields and another for magnetic fields. But it turned out that if one of these units of charge was divided by the other, then the answer was equal to the speed of light! This was thought to be an incredible coincidence. But Maxwell resolved the puzzle purely by thinking about it and what was known about electricity and magnetism, and in doing so, he revealed the true nature of light.

Maxwell brought together the equations governing electric and magnetic fields and showed that one of their solutions was to have a wave. The real thunderbolt came when Maxwell showed that these waves must travel at the speed of light. It then didn't take long to realize that this was no coincidence — the waves were light!

I'm not embarrassed to admit it — I think the theoretical calculation of the speed of light is one of the most spectacular results physics has ever had. And it's right on. As you know, light is made up of electromagnetic waves. To start calculating the speed of light with that information, you first need to examine the values typically involved with both electric and magnetic fields. The size of the force between two charges, for example, is this:

$$F = \frac{kq_1q_2}{r^2}$$

That's actually the modern shorthand version of the following equation:

$$F = \frac{q_1q_2}{4\pi\varepsilon_0 r^2}$$

where ε_0, which equals 8.85×10^{-12} C^2/(N-m^2), is a constant called the *electric permittivity of free space,* a measure of how easily an electric field passes through free space. (Sounds promising for finding the speed of light, doesn't it?)

Similarly, magnetic fields often involve the constant μ_o, the so-called *magnetic permeability of free space* (again, sounds like something you'd want to include in a calculation of the speed of light through free space). So the force between two current-carrying wires is

$$F = \frac{\mu_o I^2 L}{2\pi r}$$

And $\mu_0 = 4\pi \times 10^{-7}$ T-m/A.

So how do μ_0 and ε_0 connect to the speed of light? Well, Maxwell derived some famous equations describing how light works, and here's the payoff: He was actually able to derive the speed of light like this:

$$c = \frac{1}{\left(\mu_0\varepsilon_0\right)^{1/2}}$$

No, your eyes aren't deceiving you. Maxwell was indeed able to calculate the speed of light by simply connecting it to the two fundamental constants of electric and magnetic fields. This relation is exact, as determined by the laws of electric and magnetic fields. Now if that doesn't get you excited, I don't know what will!

Light plus friction equals hot soup

Microwave ovens provide an excellent example of how electromagnetic waves can transfer energy. Here, I use the detailed picture of the physics of the electromagnetic wave to heat a bowl of soup.

Although water molecules have no charge overall, each molecule has a positive end and a negative end because the electrons in the molecule aren't evenly distributed. Therefore, you say that the water molecule is an *electric dipole*.

When you place the polar water molecule in an electromagnetic wave, the molecule tries to align itself with the alternating electric field. This causes the molecule to rotate back and fourth as the electric field oscillates. The motion causes the water molecule to push and pull on neighboring molecules, causing them to move and vibrate — and this greater vibration of the molecules is exactly what it means for something to be at a higher temperature. The frequency of the waves in a microwave oven transfers energy to water molecules at a rate that's good for cooking: 2.45×10^9 Hz.

You've Got the Power: Determining the Energy Density of Light

Like water waves (which I touch on in Chapter 6), electromagnetic waves can carry energy. If they didn't, everyone would be in trouble, because energy from the sun would never reach Earth, meaning there wouldn't be any solar power, oil, plant life, or warmth. Not a pretty picture, huh?

To get an idea of how much energy an electromagnetic wave carries, you have to look at the wave's *energy density,* the amount of energy that wave carries per cubic meter. The units for an electromagnetic wave's energy density are joules per cubic meter.

Why not just find total energy? Well, when you're talking about a light source like the sun, you don't just switch it on and off, so you can't really think about it in terms of total energy. This section explains how you find energy density.

Finding instantaneous energy

Light waves are electromagnetic waves, so you can reasonably assume that the energy in a light wave comes from its electric and magnetic components. In an electromagnetic wave, you have an electric field and a magnetic field, which are changing with time (see the earlier section "Let There Be Light!

Generating and Receiving Electromagnetic Waves" for details). The energy in these fields is spread through the space that they occupy.

In this section, I show you how to work out the density of the energy stored in these fields at any point and any time. This gives you the foundation to figure out how much power is in an electromagnetic wave, as you see later in "Averaging light's energy density." When you know that, you can figure out things like how much energy Earth's equator receives from the sun.

Looking at electric energy density

Obviously, you need energy to set up an electric field in space. For example, to charge a capacitor (see Chapter 4), which stores energy, you have to do work to put the charges on each plate. After you've charged it, the work you did isn't lost — it's stored in the electric field between the plates. Because this field is uniform, the energy stored in it is uniformly distributed throughout the space between the two plates.

If you work out how much work you did to charge the capacitor and then divide it by the volume of the space between the plates, you have an expression for the energy density of the electric field. It turns out to be the following:

$$\text{Electric energy density} = \frac{\varepsilon_0 E^2}{2}$$

where E is the magnitude (strength) of the electric field and ε_0 is a constant equal to 8.85×10^{-12} $C^2/(N\text{-}m^2)$.

Actually, that's the amount of energy density you need to set up an electric field from any source — a parallel plate capacitor or light waves. Wherever the electric field has magnitude E, the energy density at that point is given by the preceding equation. So now you know one component of the total energy density in an electromagnetic wave: the energy density of an electric field.

Considering magnetic energy density

You can make a uniform magnetic field in a solenoid by setting up a current in the loops of wire (see Chapter 4). Setting up the current takes work, and this work is stored in the magnetic field inside the solenoid.

You can work out how much work you did to set up the uniform magnetic field and divide it by the volume of the space that it occupies to find the density of the energy stored in this field. The answer turns out to be the following:

$$\text{Magnetic energy density} = \frac{B^2}{2\mu_0}$$

where B is the magnitude (strength) of the magnetic field and μ_0 is a constant equal to $4\pi \times 10^{-7}$ T-m/A.

Guess what — that's exactly how much energy density (energy per cubic meter) you need to set up a magnetic field from any source, whether that's wires in a solenoid or an electromagnetic wave. So at any point, where the magnetic field strength is B, the density of the energy stored in the magnetic field there is given by the preceding relation.

Adding the energy densities together

Because light is made up of an electric field component and a magnetic field component, the total energy density of an electromagnetic wave is simply the sum of the two energy densities. The equation for total energy density (u) looks like this:

$$u = \frac{\varepsilon_0 E^2}{2} + \frac{B^2}{2\mu_0}$$

That's the total energy density, u, of an electromagnetic field (electric and magnetic fields together) per cubic meter. You can use this expression to work out the energy density at every point and time of the fluctuating fields of an electromagnetic wave.

Now you're making progress! So consider this question: How does nature decide which component of an electromagnetic wave to put more energy into — the electric component or the magnetic component? Turns out both components have equal energy. That is, the electric energy component is equal to the magnetic energy component, which means you can say the following:

$$\frac{\varepsilon_0 E^2}{2} = \frac{B^2}{2\mu_0}$$

That's interesting, because using that equation, along with the formula for the speed of light (from the earlier section "Calculating the speed of light theoretically"), you can do some pretty slick algebra. First, isolate E on one side of the equation:

$$\frac{\varepsilon_0 E^2}{2} = \frac{B^2}{2\mu_0}$$

$$E^2 = \frac{2B^2}{2\mu_0 \varepsilon_0}$$

$$E = \frac{B}{\left(\mu_0 \varepsilon_0\right)^{1/2}}$$

Now, because the speed of light in a vacuum is $c = \dfrac{1}{(\mu_0 \varepsilon_0)^{1/2}}$, you can just plug c into the equation:

$$E = \frac{B}{(\mu_0 \varepsilon_0)^{1/2}}$$

$$E = cB$$

So the magnitude of the electric component in a light wave is connected to the magnitude of the magnetic component by a factor of c. Well, because $\dfrac{\varepsilon_0 E^2}{2} = \dfrac{B^2}{2\mu_0}$ and $u = \dfrac{\varepsilon_0 E^2}{2} + \dfrac{B^2}{2\mu_0}$, you finally get the following equation for the energy density:

$$u = \varepsilon_0 E^2$$

Or equivalently,

$$u = \frac{B^2}{\mu_0}$$

Nice work! You've found the energy density at every point of an electromagnetic wave in terms of the strength of the electric and magnetic fields there.

Averaging light's energy density

Light's energy density depends only on electric and magnetic fields, as I show you in the preceding section. In an electromagnetic wave, these fields fluctuate.

Assume that the fields are fluctuating in the shape of a sine wave. The frequency of the fluctuations in a light wave is something like hundreds of thousands of billions of times per second (10^{14} hertz) — too fast to measure. So instead, physicists calculate the average energy density in the space occupied by an electromagnetic wave.

To get the average energy density, you work with the *root-mean-square* (rms) of the electric and magnetic fields (the maximum field divided by the square root of 2). The root-mean-square electric field is given in terms of the amplitude of the fluctuation of the sine-shaped electric field, E_0, by the following equation:

$$E_{rms} = \frac{E_0}{\sqrt{2}}$$

And for the magnetic field, it's given by

$$B_{rms} = \frac{B_0}{\sqrt{2}}$$

where B_0 is the amplitude of the sine-shaped magnetic field fluctuation. So the average energy density (u_{avg}) in a space occupied by an electromagnetic wave is $u_{avg} = \varepsilon_0 E_{rms}^2$ and $u_{avg} = \frac{B_{rms}^2}{\mu_0}$.

Here's a fun problem for you: The sun's light rays arrive with a root-mean-square E field of roughly 720 newtons per coulomb. What's their energy density?

Use the $u_{avg} = \varepsilon_0 E_{rms}^2$ equation and plug in the numbers to get

$$u_{avg} = (8.85 \times 10^{-12} \text{ C}^2/\text{N-m}^2)(720 \text{ N/C})^2 \approx 4.6 \times 10^{-6} \text{ J/m}^3$$

Looks like the time-average energy density of the sun's light rays at the Earth is 4.6×10^{-6} joules/meter^3.

Okay, now imagine a plane area, A. Suppose you want to know how much energy falls on this area every second when the electromagnetic wave is traveling straight down onto it. In a time, t, the wave will travel a distance of ct (speed times time). So all the energy that is in the volume Act will strike the plane area. You can use the average energy density formula to work out this energy:

$$u_{avg} Act = \varepsilon_0 E_{rms}^2 Act = \frac{B_{rms}^2 Act}{\mu_0}$$

So the energy falling per unit area per unit time, I, is given by

$$I = c\varepsilon_0 E_{rms}^2 = \frac{cB_{rms}^2}{\mu_0}$$

The power in a wave per unit area is the intensity (as I show you in Chapter 7). So I in the preceding formula is the intensity of an electromagnetic wave. This is just the electromagnetic equivalent of the sound intensity that you see in Chapter 7.

Using this formula, you can find the intensity of the sun's light here on Earth,

$$I = (3.00 \times 10^8 \text{ m/s})(8.85 \times 10^{-12} \text{ C}^2/\text{N-m}^2)(720 \text{ N/C})^2 = 1,380 \text{ J/s-m}^2$$

This means that every square meter of the surface of the Earth receives 1,380 joules of energy every second from the sun. (Note however, that the preceding equation only applies if the wave strikes straight down on the area, so this really applies only near the equator. Nearer the poles, you'd have to include a factor to account for the surface's tilting away from the sun at greater latitudes — obviously the North Pole doesn't receive as much energy per unit area from the sun as a Caribbean island!)

Chapter 9

Bending and Focusing Light: Refraction and Lenses

*H*ere's a cool quality of light: It interacts with matter so that it bends. Instead of just passing through the universe oblivious to everything else, light is affected by the matter through which is passes, whether that matter is dense like a diamond or thin as air.

Why does light bend? It bends because light is made up of *electromagnetic waves* — that is, tiny electric and magnetic fields — and they actually interact with the tiny electric and magnetic fields you find in matter (coming from charged particles, such as electrons and protons, and their motion).

This chapter first introduces a different way of representing light waves: the ray. Then it begins a discussion of the tricks that light can play as it bends in glass, water, and other such media. You start by getting a handle on the index of refraction, which is all about just how much light bends in any given material. You also see lenses bring images into focus, or even total internal reflection when light can't make it out of a block of glass like a prism.

Wave Hello to Rays: Drawing Light Waves More Simply

When you're exploring the various paths that light waves take as they bounce off and bend through various reflective or transparent materials, you're more interested in the places that the waves go, their directions and deflections, than the details of the wave fluctuations of electric and magnetic fields. So for simplicity, you can forget about electric and magnetic fields in most of this chapter and deal with rays of light (I point out when you need to take note of light's wavy nature). A ray just tells you the wave's direction of travel without showing the wavelength or speed or frequency or the positions of the wave peaks — the kinds of things I cover in Chapter 8.

Rays are not a new thing — you probably already think of light as rays anyway. They're just a simpler way to refer to the light wave. You can see what I mean by *rays* in terms of the light-wave picture in Figure 9-1. Here's how to interpret this figure:

- ✔ The dotted lines represent the light waves by showing the positions of the wave peaks of the electric field.

- ✔ Solid lines are some of the rays that represent the same wave. You can see that these are just lines that are always at right angles to the wave peaks — so they always lie in the direction of travel of the waves. The arrows on the rays show this direction.

In Figure 9-1a, the light waves come from a single point (that is, you have a *point source*), and they're spreading out in all directions. Because this light is traveling in all directions from the central point, any line drawn radially outward from this point is a ray.

Figure 9-1b shows another example of rays representing waves. This time you have a plane light ray traveling to the right. I've drawn three of the rays that represent this wave.

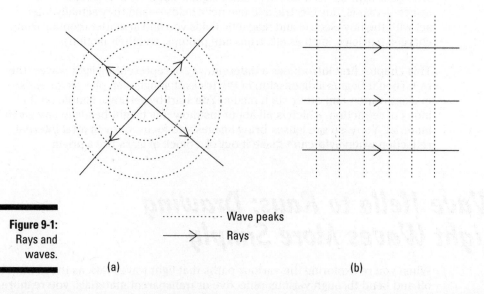

Figure 9-1:
Rays and
waves.

·········· Wave peaks
——→ Rays

(a)　　　　　　　　　　　　　　(b)

When working with light rays, just remember two basic principles:

- ✔ **Rays travel in straight lines.** When they meet a surface, they may reflect or deflect, but while they're traveling through the same medium without boundaries, rays travel in straight lines.

- ✔ **Rays are reversible.** When a light ray travels between two points (say, A to B) along a path, then the light from B to A follows the same path in the opposite direction.

Slowing Light Down: The Index of Refraction

As soon as you get past the concept that light consists of alternating E and B fields that regenerate each other, that there's a maximum speed at which light can travel (which you can calculate theoretically), and a few other items (see Chapter 8), light traveling in a straight line through a vacuum forever isn't all that interesting. Sure, you could spend some time studying the situation, but when you have it pretty well scoped out, you kind of wish something else would happen.

But when light hits and starts traveling through something else, then light becomes interesting again. When light in a vacuum enters any material and begins to travel through that material instead, the light slows down, because the electric and magnetic fields around the light in the material act as a drag. For example, when light impacts a block of transparent material (this is all theoretical, so make it a 60-pound block of diamond), the light slows down and bends.

In this section, you look at how much that material slows light down and see how much the light bends as a result. I also show you that not all light bends equally, because the index of refraction varies depending on the wavelength of the light.

Figuring out the slowdown

Light reaches its maximum speed, c, in a vacuum. That's about 3.0×10^8 meters per second, and it's all downhill from there, because whenever light travels through anything else — even air — it slows down.

The ratio of the speed of light in a vacuum, c, to the speed of light in a material, v, is a constant for any given material, and that ratio is called the *index of refraction*, n. Here's the definition of the index of refraction:

$$n = \frac{\text{speed of light in a vacuum}}{\text{speed of light in the material}} = \frac{c}{v}$$

The index of refraction is just a pure number, because it's the ratio of speeds, so it has no units, like relatively few other quantities in physics.

Generally speaking, the denser the material, the more electric and magnetic fields it has to slow light down. So diamond, for example, has a higher index of refraction than air. Table 9-1 gives a starter list of indexes of refraction for various materials. The table also includes temperature, which can affect the density of the material and therefore its index of refraction.

A material usually contracts as its temperatures decreases, so it becomes denser and its index of refraction can rise. However, water is a special case. Ice (at 0°C) has a refractive index of 1.32, and water has the higher value of 1.33. When the water freezes, the molecules form ice crystals, which happen to have a structure that's *less* dense than the original water. That's why ice floats on water.

Table 9-1	Indexes of Refraction for Various Materials	
Material	**Temperature (°C)**	**Index of Refraction (n)**
Diamond	20°C	2.42
Window glass	20°C	1.52
Benzene (liquid)	20°C	1.50
Water	20°C	1.33
Ice	0°C	1.32
Air	20°C	1.00029
Oxygen	20°C	1.00027
Hydrogen	20°C	1.00014

So if diamond has a refractive index of 2.42 at 20°C, what's the speed of light in a diamond? Well, it's

$$v_{\text{diamond}} = \frac{c}{n_{\text{diamond}}} \approx \frac{3.0 \times 10^8 \text{ m/s}}{2.42} \approx 1.2 \times 10^8 \text{ m/s}$$

So light travels at only 1.2×10^8 meters per second in diamond. Positively pokey.

Calculating the bending: Snell's law

When light slows, it bends. You can put the index of refraction to work with Snell's law, which tells you exactly how much light bends when it enters a new medium. (See the preceding section for info on the index of refraction.)

The incident (incoming) light comes in at an angle of θ_1, measured with respect to a line perpendicular to the material's surface — that perpendicular line is called a *normal* (see Figure 9-2). And when the light bends and travels off into the medium, it goes at a new angle with respect to the normal, θ_2. Here's how the angles relate:

$$n_1 \sin \theta_1 = n_2 \sin \theta_2$$

where n_1 is the index of refraction of the medium the light is coming from (it doesn't need to be vacuum to have Snell's law work) and n_2 is the index of refraction of the medium the light enters (which could be diamond, or glass, or even vacuum).

Note that if you know the incident angle and the indexes of refraction of the materials involved, you can solve for the angle at which light heads off into the new medium like this:

$$\sin^{-1}\left(\frac{n_1 \sin \theta_1}{n_2}\right) = \theta_2$$

That's a nice result — it tells you what angle you can expect light to go speeding off at in a new medium.

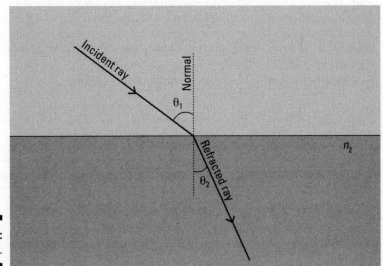

Figure 9-2:
Snell's law.

Here's one thing that Snell's law tells you that's not immediately obvious: If you're going from less-dense to denser material, light is bent toward the normal; if you're going from denser to less-dense material, light is bent away from the normal.

For instance, if you look at Figure 9-2, you see the light ray traveling from less-dense to denser material, and the light bends toward the normal. Now if you remember that light rays are reversible, you can imagine the same ray going in the opposite direction (from the denser to the less-dense material) — then the ray bends away from the normal.

Now try some numbers. Say that light goes from air (which you can treat as vacuum for the purposes of this example) to your 60-pound diamond block. And say that light hits the diamond at 65° with respect to the normal. What's the angle the light bends to inside the diamond?

That is, you have $n_1 = 1.00$, $n_2 = 2.42$, and $\theta_1 = 65°$, and you need to find θ_2. You can use Snell's law like this:

$$\theta_2 = \sin^{-1}\left(\frac{n_1 \sin\theta_1}{n_2}\right) = \sin^{-1}\left(\frac{1.00 \sin 65°}{2.42}\right) \approx 22°$$

So the light comes in at 65° and ends up at 22°.

Rainbows: Separating wavelengths

Here's something that you may not like to hear because it complicates things a bit: The index of refraction of materials varies slightly depending on the wavelength of the light. On the other hand, you probably like the results of this fact: rainbows. Because the colors in sunlight (which contains all colors) bend different amounts in water droplets, you get a separation of colors into that familiar display of rainbows.

The index of refraction does vary by light wavelength but not strongly (so physicists often ignore it). Table 9-2 lists some values for the various colors of light and what the corresponding indexes of refraction are in glass.

Table 9-2	Indexes of Refraction According to Wavelength	
Color	*Wavelength (nanometers)*	*Index of Refraction in Glass*
Red	660	1.520
Orange	610	1.522
Yellow	580	1.523

Color	Wavelength (nanometers)	Index of Refraction in Glass
Green	550	1.526
Blue	470	1.531
Violet	410	1.536

Say that light is incident at 45° to a sheet of glass. How much does red light (λ = 660 nm) bend, compared to violet light (λ = 410 nm)?

Snell's law tells you that $n_1 \sin \theta_1 = n_2 \sin \theta_2$, so

$$\sin^{-1}\left(\frac{n_1 \sin \theta_1}{n_2}\right) = \theta_2$$

The light is first traveling through the air, so n_1 = 1.00. For red light in glass, the index of refraction is 1.520, so you get

$$\theta_2 = \sin^{-1}\left(\frac{1.00 \sin 45°}{1.520}\right) \approx 27.7°$$

For violet light, the index of refraction in glass is 1.536, so you get

$$\theta_2 = \sin^{-1}\left(\frac{1.00 \sin 45°}{1.536}\right) \approx 27.4°$$

So as you can see, you get different amounts of bending depending on the color of light. Note that the angle calculated is with respect to the normal, so violet light is bent slightly more than red light here.

Because of the differing indexes of refraction for different wavelengths, light splits in a prism (see Figure 9-3). Here's how it works: when light enters the prism, it's going from air into a medium with a higher index of refraction — typically glass — so it bends in the glass toward the *normal* (a line perpendicular to the surface). Because the index of refraction is stronger for shorter wavelengths, red light (with a longer wavelength) bends less than violet light (with a shorter wavelength). When the light emerges from the prism, it bends away from the normal, and how much it bends depends on the index of refraction — so red light is further separated from violet light.

In actual rainbows, the light not only refracts when it enters the water droplet but also reflects inside the water droplet. You can read more about this phenomenon in the sidebar "Reflecting on rainbows," later in this chapter.

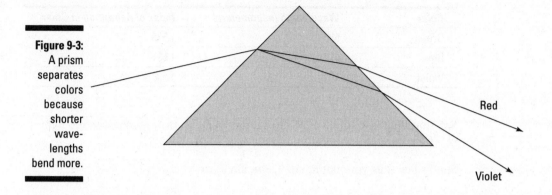

Figure 9-3:
A prism separates colors because shorter wavelengths bend more.

Red

Violet

Bending Light to Get Internal Reflection

When light enters a material with a lower index of refraction, that light bends away from the *normal* (an imaginary line perpendicular to the material's surface). If the incident light comes in at a large enough angle, the light may bend away so much that it doesn't refract at all — it gets reflected instead.

In this section, I discuss two cases in which you get reflection. In the first, all the incident light is reflected. In the second, only *polarized light* — light with aligned electric and magnetic fields — is reflected, and the rest of the light is refracted as it enters the less-dense material.

Right back at you: Total internal reflection

Sometimes light doesn't make it out of a material, and it ends up bouncing around inside. Perhaps you've noticed that when you turn a glass paperweight, one of the internal edges sometimes looks like a mirror, reflecting with a silvery appearance. That's total internal reflection.

To see how this works, take a look at Figure 9-4. Light is going from a dense medium such as glass to air. That means that the light bends away from the normal when it gets into the air, as you see in ray 1. If you keep increasing θ_1, ultimately, θ_2 reaches 90° — that is, the light just skims along the glass surface, as in ray 2. If you increase θ_1 any more, the light will be reflected back into the glass, as you see in ray 3. This is what's known as *total internal reflection*. When light goes from a dense medium to a less-dense medium, it bends away from the normal, and if the incident angle becomes large enough, the light will be reflected back into the denser medium where the two materials meet.

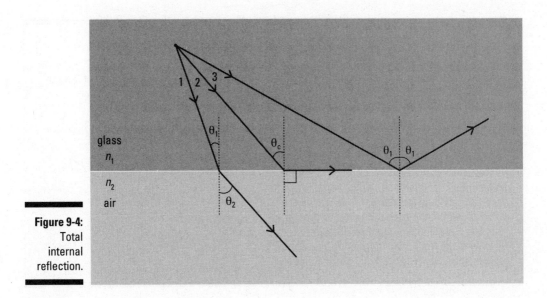

Figure 9-4:
Total
internal
reflection.

Total internal reflection happens when the angle at which the light tries to exit the dense medium, θ_2, becomes so large that it reaches 90°. Right at that point, when the light ends up skimming the glass/air interface, you have total internal reflection. The incident angle at which this happens is called the *critical angle* — θ_c. At the critical angle, light ends up with an exit angle of 90° with respect to the normal. In other words, when $\theta_1 = \theta_c$, $\theta_2 = 90°$.

What value does θ_c have? You can use Snell's law to find out:

$$n_1 \sin \theta_1 = n_2 \sin \theta_2$$

Plugging in the values, you get

$$n_1 \sin \theta_c = n_2 \sin 90°$$

$$\sin \theta_c = \frac{n_2 \sin 90°}{n_1}$$

Because sin 90° = 1, you have the following for the critical angle for total internal reflection:

$$\sin \theta_c = \frac{n_2}{n_1}$$

$$\theta_c = \sin^{-1}\left(\frac{n_2}{n_1}\right)$$

Note that this equation requires that $n_2 < n_1$ (otherwise you can't have total internal reflection).

Reflecting on rainbows

Rainbows form when water droplets are in the air. Creating rainbows is a matter of varying indexes of refraction. Red light, for instance, bends less when it enters the water droplet, and violet light bends more.

Specifically, sunlight enters the droplet, and all the colors of the rainbow begin to separate immediately. In this case, violet bends more than red. Then, given the angle of incidence when the split light tries to leave the droplet, total internal reflection occurs. The light ends up leaving the front of the droplet — and violet is bent more than red again, as this figure

shows. The result — all the colors of the rainbow. Physics does it again!

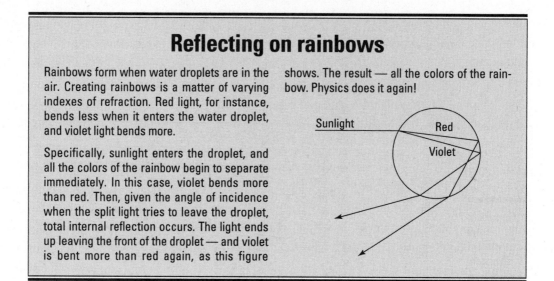

For example, say you have a diamond ring and that light is bouncing around inside the gem. What's the critical angle beyond which light already inside the diamond gets totally internally reflected back into the diamond? You can use the equation for total internal reflection. The index of refraction of air is near 1.00, and the index of refraction for diamonds is 2.42, so you have the following:

$$\theta_c = \sin^{-1}\left(\frac{n_1}{n_2}\right) = \sin^{-1}\left(\frac{1.00}{2.42}\right) = \sin^{-1}(0.413) \approx 24.4°$$

So if light hits the diamond-air interface at an angle of more than 24.4°, it'll bounce back into the diamond, and that facet of the diamond acts like a mirror — that's one of the reasons properly cut diamonds seem to exhibit so much fire.

Polarized light: Getting a partial reflection

Here's a peculiar fact about light — when it bounces off a nonmetallic surface, it gets polarized. That means that the light rays' electric fields are lined up and its magnetic fields are lined up.

When you talk about polarization, you normally discuss the direction of the electric (*E*) field in the light ray. The *E* field oscillates in a direction that's perpendicular to the light ray's direction of travel, and the plane that the light ray and the *E* vector form is called its *polarization*. So if you have a light ray coming toward you and its *E* vector is oscillating horizontally, the light is polarized horizontally.

With normal light, the *E* vector can oscillate in any direction perpendicular to the direction of travel. But when you bounce light off a nonmetallic surface, the reflected light ends up being polarized to some extent in the plane of the surface. For example, if light bounces off a pool of water, the reflected light ends up being chiefly polarized in the horizontal direction.

In Figure 9-5, you see the incident ray coming in from the left. A number of arrows in all directions (perpendicular to the direction of travel, of course) represent the unpolarized electric-field component of this ray. The reflected ray on the right is totally polarized, so it has electric field oscillations only in the horizontal direction. The refracted ray is partially polarized, because the reflected wave preferentially carried away electric field oscillations in the horizontal direction, leaving the refracted wave with relatively few. You can see this in Figure 9-5 — in the refracted ray, the arrows representing the horizontal electric field oscillations are diminished compared to the others.

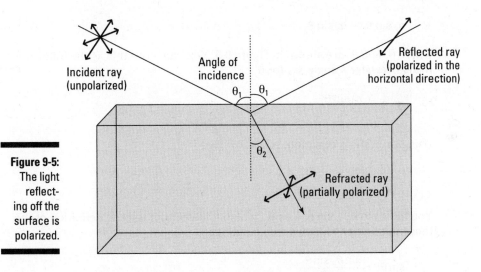

Figure 9-5:
The light reflecting off the surface is polarized.

Reflecting polarized light at Brewster's angle

When light bounces off a nonmetallic surface, the amount of polarization depends on the angle of incidence (with respect to the normal, an imaginary line perpendicular to the surface). And at an angle of incidence called *Brewster's angle,* θ_B, the polarization is total. So when light reflects off a pool of water at Brewster's angle, the reflected light is completely polarized in the horizontal direction. Here's the formula for Brewster's angle:

$$\tan\theta_B = \frac{n_2}{n_1}$$

where n_1 and n_2 are the indexes of refraction. So what is Brewster's angle for water? Well, the index of refraction for water is about 1.33 and the index of refraction for air is about 1.00, so you have the following:

$$\tan\theta_B = \frac{1.33}{1.00}$$

$$\theta_B = \tan^{-1}1.33 \approx 53°$$

So Brewster's angle for water is 53°.

Noting the angle between the reflected and the refracted rays

You can prove that the refracted ray, which enters the water, is at 90° with respect to the reflected ray if the incident light enters the water at Brewster's angle (the angle at which polarization is total). To prove this, start with Snell's law:

$$n_1 \sin\theta_1 = n_2 \sin\theta_2$$

where θ_1 and θ_2 are shown in Figure 9-5. You can write this as the following, using Brewster's angle, θ_B, for θ_1:

$$\sin\theta_B = \frac{n_2 \sin\theta_2}{n_1}$$

Using Brewster's equation, you know that

$$\tan\theta_B = \frac{n_2}{n_1}$$

You have tan θ_B = sin θ_B/cos θ_B, which follows from the trig definition of the tangent, so do the following calculations:

$$\sin\theta_B = \frac{\sin\theta_B \sin\theta_2}{\cos\theta_B}$$

$$1 = \frac{\sin\theta_2}{\cos\theta_B}$$

$$\cos\theta_B = \sin\theta_2$$

And because sin θ = cos(90° – θ), you can say that

$$\cos\theta_B = \cos(90° - \theta_2)$$

$$\theta_B = 90° - \theta_2$$

So Brewster's angle is 90° from the refracted angle. Cool.

Cutting the glare with Polaroid sunglasses

Polaroid sunglasses take advantage of the fact that light is polarized when it bounces off flat surfaces. These sunglasses are created with thousands of elongated crystals that are applied to a film, which is then stretched. Stretching the film aligns the crystals, which then permit only light of a certain polarization — parallel to the crystals — through. That's why people sometimes wear Polaroid sunglasses when fishing: The sunglasses filter out the light bouncing off the water, which would otherwise be glaring. A nice effect, isn't it?

Getting Visual: Creating Images with Lenses

Many weird and wonderful things happen to light when it hits curved surfaces, so in this section, you step into a world of images. You find out what the field of optics considers to be an *object* and how images are made by the curved surfaces of lenses. This is the physics that led to telescopes and microscopes, which opened up new doors of perception of the universe.

Here, I demonstrate how you work out where and how big an image will be simply by drawing a few lines. Such ray drawings can give you a good mental picture of what lenses do to the light that passes through them. After this, I show you a couple of equations, which tell you exactly where the images are and how big they are, without your having to take out your ruler and pencil.

Defining objects and images

As far as the field of optics is concerned, an *object* is simply a source of light rays. It doesn't have to glow with light; it can just reflect light from another source. The important point is that light rays should radiate away from the object. For example, this book would be considered an object in physics. The book isn't generating the light, only reflecting it from your lamp or the sun or whatever the light source is where you're reading.

For simplicity, this book considers only very simple objects like *point sources*, which are simply points that radiate rays, or line sources. A *line source* is simply a line that radiates rays in all directions from every part of it — physicists draw these as arrows, because sometimes you find them upside down, and the arrowhead emphasizes direction.

So much for objects — now what about the images that are made from them? Well, the simplest example of an image is one you probably see every day — your own image in the mirror. In this case, you're the object (no offense!), and the mirror reflects the rays coming from you and makes an image behind the mirror. The *image* is the point from which the rays leaving your face *seem* to be coming from. You know that you — the object — are not behind the mirror, but your image appears behind. This kind of image is called a *virtual image*.

Another example of an image is the one projected in a cinema. Each frame of the movie is turned into an image, which is projected onto the screen. This is a different type of image from the one you see in the mirror, because the rays don't just *appear* to be coming from the image — they actually converge onto the image. That is, the rays make the picture all come together on the movie screen. Because this type of image can be made to fall on a screen, it's called a *real image*.

Now it's coming into focus: Concave and convex lenses

You can find lenses everywhere — in standard digital cameras, in TV cameras, perched on the end of peoples' noses, in flashlights, even sometimes in peoples' watches when the really tiny date has to be magnified. A lens is simply a transparent object (usually a disk of glass), that takes an object and makes an image. It can do this because it has two curved surfaces.

Here are two types of lenses:

- ✔ **Convex (converging):** A *convex lens* curves so that it bulges in the middle (see Figure 9-6). When you put a point object in front of this lens, some of the rays that radiate away travel through the lens. When they meet the first surface, they refract, or bend, and when they leave the lens, they refract some more. The effect of the convex lens in Figure 9-6a is to gather all these rays together back to a point — this is your image. Because all the rays converge there, it's a *real image*.

  If a number of parallel rays strike the lens, then they all converge on a point called the *focal point,* as you see in Figure 9-6b. You may have already discovered the focal point for yourself if you've ever tried to focus sunlight into a single bright spot with a magnifying glass. The sun is so far away that the rays coming from it are pretty much parallel, so when they pass through your magnifying glass, they converge at the focal point. If you place a piece of paper there, then all the converging rays can make the paper burn.

✔ **Concave (diverging):** A *concave lens* is narrower in the middle. This time, when the rays from the object strike the lens, they diverge. All the rays appear to diverge from a certain point, and this is the *virtual image.* See Figure 9-7a.

Now if you send a bunch of parallel rays into a concave lens, they all diverge, but they all appear to be diverging from the *focal point* (see Figure 9-7b).

The way I remember the difference between convex lenses and concave lenses is that a diverging lens forms a sort of a cave (because its middle is all hollowed out), as in con*cave.*

The distance between the lens and the focal point is called the *focal length, f.* The strength of a lens is measured solely by this length — if it's shorter, then the rays are bent through a larger angle, and the lens is stronger.

In many lenses, one of the lens's surfaces may curve more than the other. Even in those cases, there's still only one focal length, so the focal point on each side of the lens is at the same distance, *f.* Also, however the sides are curved, if the lens is thicker in the middle than at the edges, it's convex; otherwise, it's concave.

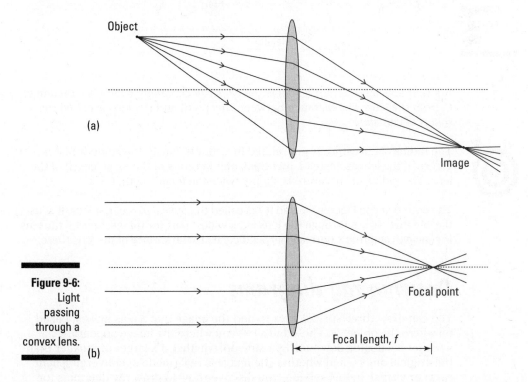

Object

(a)

Image

Figure 9-6:
Light
passing
through a
convex lens.

Focal point

Focal length, *f*

(b)

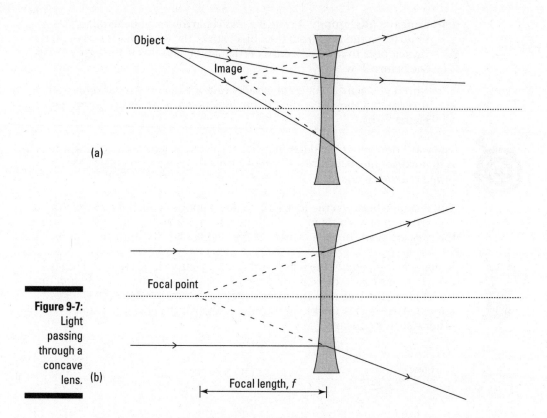

Figure 9-7:
Light
passing
through a
concave
lens.

Another special point for a lens is the *center of curvature*, which is a distance, *C*, from the lens. The distance between this point and the lens is called the *radius of curvature*.

The radius of curvature is not related in a simple way to the amount of curvature of the lenses; instead, just think of it in terms of the focal length of the lens. The radius of curvature is simply twice the focal length, $C = 2f$.

The dotted line in Figures 9-6 and 9-7 is called the *optical axis* of the lens. It's just the line that passes through the lens at its widest part (or thinnest part if the lens is concave) and that's normal (perpendicular) to the surface of the lens there.

Drawing ray diagrams

You can draw three special lines to find the image that a lens makes, based on where the object is. These lines tell you where the image appears, whether it's upside down or right-side-up, whether it's larger or smaller than the original object, and whether the image is real (made of converging light rays) or virtual. In this section, you discover how to draw ray diagrams for both convex (converging) and concave (diverging) lenses.

The object I use in each section is a *line object,* which looks like an arrow. The arrowhead makes sure you always know whether the image is upside down or the right way up. You can think of this type of object as just being a lot of point objects all in a line. (For more on objects and point and line sources, see the earlier section "Defining objects and images.")

X marks the spot: Finding images from convex lenses

The position and size of the image depend on the position and size of the object. For a convex (converging) lens, here's how to draw three special lines that help you figure out the position and size of the image (see Figure 9-8):

- ✔ **Ray 1:** One ray leaves the object, travels toward the center of the lens, and travels straight through without being deflected at all.

- ✔ **Ray 2:** Another ray travels from the object, parallel to the axis of the lens, and is deflected so that it passes through the lens's focal point.

- ✔ **Ray 3:** A third ray travels from the object and passes through the focal point on the near side before reaching the lens. The lens deflects this ray so that it then travels parallel to the axis of the lens.

The image is located where these three lines cross.

You can draw these ray lines for any point on your object and find every point of the image, but for simplicity, most people draw only the lines from the tip of the object (the point of the arrow). And although I draw three rays in Figure 9-8, you really need only two rays to locate the image. Three is better for safety — as a check on the other two — but if you know what you're doing and are under time pressure (such as when you're taking a test), two is enough.

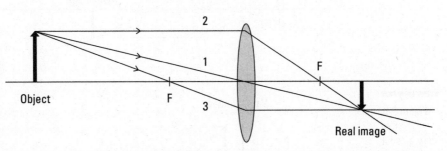

Figure 9-8: Drawing three special rays for a convex lens.

If you go through the line-drawing process for convex lenses, you encounter three special cases for what the image looks like. Here are these cases, which are all based on the position of the object (if you need more info on the focus and radius of curvature, see the earlier section "Now it's coming into focus: Concave and convex lenses"):

✔ **The object is beyond the radius of curvature, *C:*** If the object is this far out, the image is real, upside down, and smaller than the object (see Figure 9-9a).

✔ **The object is between the radius of curvature, *C*, and the focal length, *f:*** Here the image is still real and upside down, but now it's larger than the object (see Figure 9-9b).

When the object is at the center of curvature, then its image is the same size as the object, and the closer the object gets to the focal point without crossing, the larger the image becomes.

✔ **The object is closer to the lens than the focal length, *f:*** This is an interesting case, because for the first time, you don't get a real image (see Figure 9-9c). You can't even draw the third ray, because it doesn't go through the lens. There's no place in space that the three rays come together, no place that you can bring an actual physical screen and get an in-focus image — the image is virtual. It's also larger than the object and right-side-up.

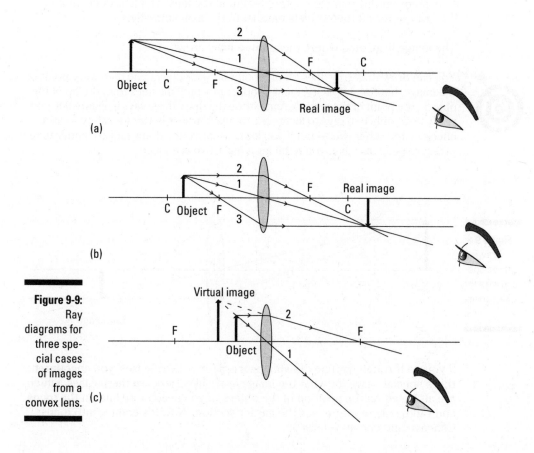

(a)

(b)

Figure 9-9:
Ray
diagrams for
three spe-
cial cases
of images
from a
convex lens.
(c)

Recognize the situation in which the object is between the focal point and the lens? That's the case where a converging lens forms an upright image that's larger than the object you're looking at, on the same side of the lens as the object itself (the image is virtual, so no light rays actually come together to form the image, but looking through the lens bends the rays so that they appear to be coming from the image). That's a *magnifying glass* — and the fact that you don't get an enlarged upright image until the object is between the focal point and the lens shows why you have to hold the lens up close to whatever you're trying to magnify. Cool, eh? Sherlock Holmes would be proud.

Going virtual with concave lenses

With a concave (diverging) lens, light bends away from the horizontal after passing through the lens. You can see a diverging lens in Figure 9-10.

So when you have a diverging lens, can you work with ray diagrams to find the image? Absolutely — but this time you use only two rays:

✔ **Ray 1:** This ray goes from the tip of the object and through the center of the lens. Figure 9-10 shows that this ray goes through the center of the lens and isn't deflected at all. Easy.

✔ **Ray 2:** This ray goes horizontally from the tip of the object to the lens, parallel to the axis of the lens; then the ray deflects away from the axis along a line that passes through the nearest focal point — that is, the ray travels as though it came from the nearest focal point. Figure 9-10 shows that the second ray bends *away* from the horizontal on the other side of the lens.

Figure 9-10:
A concave lens produces a smaller, upright virtual image.

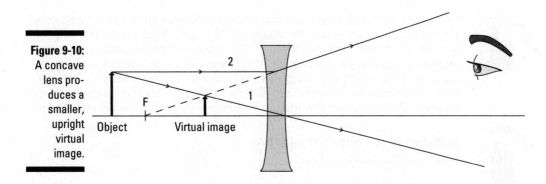

So if the second ray is bent away from the horizontal, where does it intersect the first ray? The answer is that they don't intersect on the side of the lens that the observer (who is looking through the lens) is on. Instead, you extend the rays back through the lens until they intersect. Because you're extending these rays in a straight line back to somewhere they don't actually exist, the image is *virtual*. In other words, the image forms on the same side of the lens as the object, as you see in Figure 9-10.

Regardless of where the object is, the virtual image from a concave lens is always right-side-up, and it's no farther from the lens than the focal length. The image is also upright and smaller than the object.

Getting Numeric: Finding Distances and Magnification

With a few lens equations, you can find out where images appear and how big they are. Drawing ray diagrams (as I show you in a preceding section) is a good way to get a strong picture of what lenses do, but when you have that in mind, you'll find these equations a much quicker way of finding out what your lenses are doing.

There's really nothing mystical about the equations in this section — they just come from the laws of refraction (which you can examine in "Slowing Light Down: The Index of Refraction," earlier in this chapter). People derived these equations by applying the law of refraction to the curved surfaces of the lenses, but you don't have to bother doing that — here, I just show you how the equations work.

Going the distance with the thin-lens equation

Using the thin-lens equation, you can relate the distance an object is from a lens, the distance from the lens to the image, and the lens's focal length. The equation is called the *thin-lens equation* because it's actually an approximation, and that approximation really holds only for "thin" lenses — that is, lenses whose bending power isn't too great (stronger lenses have a shorter focal length, so strong lenses have to be more curved and therefore thicker). This section gives you the equation, shows you how it works, and provides a couple of example calculations.

Introducing the thin-lens equation

Here's the thin-lens equation:

$$\frac{1}{d_o} + \frac{1}{d_i} = \frac{1}{f}$$

This equation relates the distance the object is from the lens (d_o) and the distance the image is from the lens (d_i) with the focal length, f.

The signs of d_o, d_i, and f are important. For instance, you give converging lenses a positive focal length f, but diverging lenses get a negative focal length (that's because the image forms on the other side of the lens from the observer). And if you get a negative distance for the image distance, d_i, that means the image is virtual (which, for a lens, means that the image forms on the same side as the object).

The best way to state the rules for the signs of d_o, d_i, and f in the thin-lens equation is in terms of the incoming and outgoing sides of the lens (see Figure 9-11). When light from an object travels through a lens, I call the side that the light enters the *incoming side;* the side of the lens through which the light leaves is the *outgoing* side. Then the rules are simple to state:

- ✔ **Object distance, d_o:** When the object is on the incoming side of the lens, then the distance from the object to the lens is positive (in this book, this is always the case).

- ✔ **Image distance, d_i:** When the image is on the outgoing side of the lens, then the image distance is positive; otherwise, it's negative.

- ✔ **Focal length, f:** When the lens is convex, its focal length is positive; otherwise, it's negative.

Note: This book always pictures the object to the left of the lens, so the incoming side is on the left in the figures (including Figure 9-11). But these rules still apply in exactly the same way if this situation is reversed, because then the incoming and outgoing sides would also reverse.

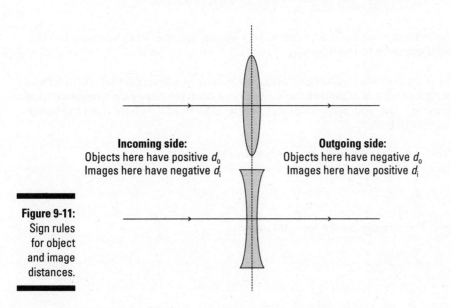

Incoming side:
Objects here have positive d_o
Images here have negative d_i

Outgoing side:
Objects here have negative d_o
Images here have positive d_i

Figure 9-11: Sign rules for object and image distances.

Doing calculations with the thin-lens equation

Now try some numbers with the thin-lens equation. Say that you have a camera with a converging lens that has a focal length of 5.0 centimeters, and the flower you're taking a picture of is 2.00 meters in front of the lens. How far on the other side of the lens does the image form? You can put the thin-lens equation to work right away:

$$\frac{1}{d_o} + \frac{1}{d_i} = \frac{1}{f}$$

Rearranging this gives you

$$\frac{1}{d_i} = \frac{1}{f} - \frac{1}{d_o}$$

Combining the fractions and solving for d_i gives you

$$d_i = \frac{1}{\frac{d_o - f}{f d_o}}$$

$$d_i = \frac{f d_o}{d_o - f}$$

So plugging in the numbers gives you the following:

$$d_i = \frac{(0.050 \text{ m})(2.00 \text{ m})}{2.00 \text{ m} - 0.05 \text{ m}} = \frac{0.10 \text{ m}}{1.95} \approx 0.051 \text{ m} = 5.1 \text{ cm}$$

So the image forms at 5.1 centimeters behind the lens of the camera (on the side opposite to the flower).

Now try one with a diverging lens. Say you have a diverging lens with a focal length of –5.0 centimeters (See? I told you diverging lenses get negative focal lengths), and you place an object 7.0 cm in front of it. Where does the image appear to form?

You can use the thin-lens equation like this:

$$\frac{1}{d_o} + \frac{1}{d_i} = \frac{1}{f}$$

And I've already solved for d_i this way:

$$d_i = \frac{f d_o}{d_o - f}$$

Plugging in the numbers gives you the answer:

$$d_i = \frac{(-0.050 \text{ m})(0.070 \text{ m})}{0.070 \text{ m} - (-0.050 \text{ m})} = -\frac{0.0035 \text{ m}^2}{0.12 \text{ m}} \approx -0.029 \text{ m} = -2.9 \text{ cm}$$

So the image forms at –2.9 centimeters — that is, between the focal length and the lens. Note that this result is negative. That means that the image is visible only by looking through the lens — not on the same side of the lens as the observer. In other words, it's a virtual image, as you'd expect from the type of lens and the placement of the object (to see why, check out the ray diagram in Figure 9-10).

Sizing up the magnification equation

The thin-lens equation tells you where an image will form, but it doesn't tell you very much about the image itself. Sometimes, you want to know whether that image is bigger or smaller than the object and whether it's upright or upside down with respect to the object. That's where the magnification equation comes in.

Finding the magnification equation

Say that h_i is the height of the image and h_o is the height of the object. You can see that the magnification of the image compared to the object would be m, like this:

$$m = \frac{h_i}{h_o}$$

Figure 9-12 shows two of the rays that go to make the image of an object from a convex lens. The figure shows the size of the object and the image, along with the distances of the object and image from the lens. The ray that travels straight through the lens makes an angle θ with the axis. By using geometry and similar triangles (shaded gray in Figure 9-12), you can show that the magnification is equal to the ratio of the image distance to the object distance like this:

$$m = \frac{h_i}{h_o} = -\frac{d_i}{d_o}$$

Note that a negative value for the magnifications means the image is upside down with respect to the object, and positive magnification (the kind magnifying glasses and most optical telescopes produce) means that the image is right-side-up compared to the object.

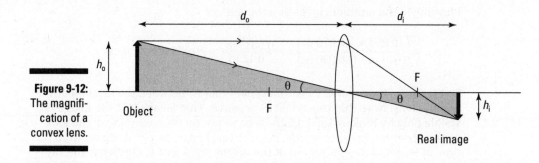

Figure 9-12:
The magnifi-
cation of a
convex lens.

h_o

Object

d_o

θ

F

F

θ

d_i

θ

h_i

Real image

TIP

Why does the magnification equation have a minus sign in it? That's because the magnification of a magnifying glass — that is, a converging lens — is considered positive when the image is virtual, because you look through the lens to see the image and it's right-side-up. But because the image is virtual, the distance to the image from the lens, d_i, is negative. The minus sign in the magnification equation corrects that — so even though the image comes out virtual (which means negative in the thin-lens equation), it's still upright (which by convention means positive in the magnification equation).

Plugging in some numbers

Now you can figure out the distance from a lens to an image using the thin-lens equation, and because you already know the distance from the object to the lens, you can use the magnification equation to figure out the magnification.

Try some numbers. Start by taking a look at the converging-lens problem from the earlier section "Going the distance with the thin-lens equation": The camera has a converging lens that has a focal length of 5.0 centimeters, and the object you're taking a picture of is 2.0 meters in front of the lens. In this problem, the distance from the lens to the image turns out to be 5.1 centimeters.

What's the magnification of the converging lens in this setup? You can use the magnification equation:

$$m = -\frac{d_i}{d_o}$$

Plugging in the numbers gives you the answer:

$$m = -\frac{5.1\ \text{cm}}{200\ \text{cm}} \approx -2.6 \times 10^{-2}$$

So the magnification is -2.6×10^{-2}. That tells you a few things. First, note that magnification is another of those relatively few quantities in physics that has no units — it's just a multiplier. Second, the negative sign tells you that the image formed is upside down with respect to the object (did you know that camera images are inverted?). And third, it tells you that the magnification is very small, so you can capture big objects on small film surfaces or pixel arrays (of digital cameras).

Now take a look at the diverging lens. In the second problem I solve in "Going the distance with the thin-lens equation," the focal length is –5.0 centimeters, and the object is placed 7.0 centimeters in front of it. The image forms at –2.9 centimeters. So what's the magnification of the lens with this setup? You can use the magnification equation:

$$m = -\frac{d_i}{d_o}$$

Plugging in the numbers here gives you this:

$$m = -\frac{(-2.9 \text{ cm})}{7.0 \text{ cm}} \approx 0.41$$

So although the image is upright, it's still smaller than the object (putting it closer to the lens will result in magnification greater than 1).

Combining Lenses for More Magnification Power

You can use lenses together — in fact, that's one of their most popular uses, in microscopes and telescopes and such. Lenses used in combination are almost always converging lenses. In this section, you look at how combining two lenses gives you more magnification power, and you see how such combinations usually work.

Understanding how microscopes and telescopes work

When you combine two converging lenses, the first lens is nearest the object, so it's called the *objective lens*. As you can see in Figure 9-13, the object is farther away from the objective lens than the focal length (f_o) of the objective lens. In a microscope, the object you're looking at may not be much beyond the focal length, but in a telescope, the object is always much farther from the objective lens.

As you can see in the figure, rays from the tip of the object form an image past the focal length of the objective lens on the right side of the lens. This image is larger than the object, and it's inverted — it's also a real image.

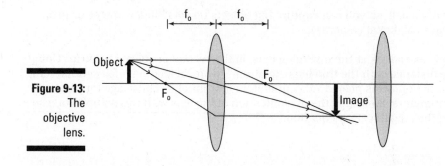

Figure 9-13:
The objective lens.

Seeing clearly with corrective lenses

Two common vision problems are *nearsightedness,* where people can focus only on near objects, and *farsightedness,* where people can focus only on objects farther away. Corrective lenses help with both — diverging lenses for nearsightedness and converging lenses for farsightedness.

The diagram at the top of the following figure shows an uncorrected nearsighted eye. The problem here is that the lens of the eye tends to focus objects some distance in front of the retina. As a result, objects look blurry. The bottom diagram shows the same eye corrected with a diverging lens. Now the diverging lens makes the rays from objects diverge slightly to counteract the overly strong converging powers of the lens of the eye. As a result, the image comes into focus directly on the retina, as it should.

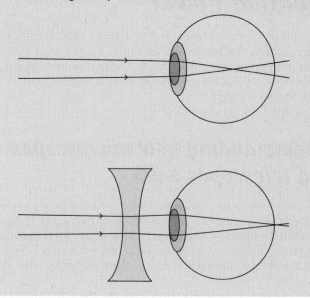

The diagram at the top of the next figure shows an uncorrected farsighted eye. Here, the problem is that the lens of the eye doesn't converge light passing through it enough. As a result, the image forms past the retina. The solution is to use a converging lens, as in the bottom diagram. The converging lens makes the light rays from the object converge a little, which helps the lens of the eye focus the image, which appears on the retina.

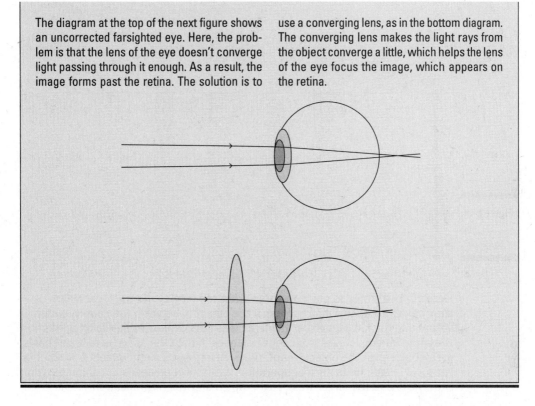

Now here's the clever part — the image from the objective lens becomes the *object* for the second lens. That is, the second lens looks at the image (which is real) just as though it were an actual object. Because light rays come together at the image, this works very well.

The second lens is called the *eyepiece* (not surprisingly, because that's the lens that is closest to the eye). Everything is set up so that the first image, the one created by the objective lens, falls just inside the eyepiece's focal length, f_e. That ensures that the second lens magnifies the image to a very large size. You can see this in Figure 9-14.

This time, the eyepiece creates a virtual image (so you can see it by looking through the eyepiece, as with microscopes and telescopes). Thus, the final image ends up large and inverted, as you can see in Figure 9-14. But how large? What's the magnification of such a combination of lenses?

Well, the image from the objective lens is magnified. Then this magnified image is the object for the eyepiece lens, which magnifies again. The total resulting magnification is the product of the magnification from each lens.

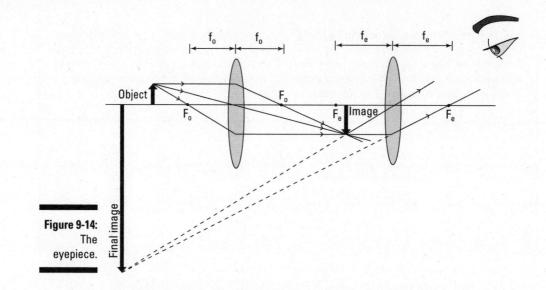

Figure 9-14:
The
eyepiece.

Here's why it's best to place the object not too far beyond the focal length of the objective lens in a microscope: If the object is between the center of curvature and the focal length of a convex lens (as I explain earlier in "X marks the spot: Finding images from convex lenses"), then the image is real and has a greater size than the object — and this size increases as the object approaches the focal length. In its normal operation, a microscope needs a real image from the object, and the larger this image, the greater the magnification.

Getting a new angle on magnification

Magnification for microscopes and for telescopes is often figured in terms of *angular magnification*. That is, the object you want to look at takes up a certain angle of your vision (for example, the moon takes up about half of a degree of the 360° you can see by turning completely around), and if you use a telescope, the object looks larger (the moon may take up what seems like three times the same angle). The symbol for angular magnification is *M*. In this section, you use some formulas to find the magnification of microscopes and telescopes.

Getting up close and personal with microscopes

Microscopes are typically made from two converging lenses in combination. Here, the object is between one and two focal lengths from the objective lens. If the distance between the objective lens and the eyepiece lens is *L*, then the angular magnification for a microscope turns out to be the following:

$$M = \left(\frac{L - f_e}{f_o f_e} \right) N$$

N is the distance to the near point for the eye. The *near point* is the closest you can hold, for example, some text and still read it. For a normal eye, N equals 25 centimeters.

Suppose you have an eyepiece with focal length of 5.0 centimeters and an objective lens with a focal length of 0.40 centimeters. The length between the two lenses, L, is 25.0 centimeters. What is the angular magnification of the microscope?

Plugging in the numbers — using $N = 25.0$ centimeters — gives you the answer:

$$M = \left(\frac{L - f_e}{f_o f_e}\right) N = \left(\frac{25.0 \text{ cm} - 5.0 \text{ cm}}{(0.40 \text{ cm})(5.0 \text{ cm})}\right)(25.0 \text{ cm}) = 250$$

So the angular magnification of the microscope is 250.

Bringing the heavens near with telescopes

Like microscopes, optical telescopes are frequently made with two converging lenses. With telescopes, the object you're looking at is a far distance away compared with the distance to the eye's near point, N, and the focal length of the objective lens.

In this case, you can make some approximations, and the angular magnification of a telescope is about equal to the following:

$$M \approx -\frac{f_o}{f_e}$$

Say, for example, that you have a telescope whose objective lens has a focal length of 100 centimeters and an eyepiece with the focal length of 0.5 centimeters. What angular magnification will the telescope give you?

You can use the angular-magnification equation and plug in numbers like this:

$$M \approx -\frac{f_o}{f_e} = -\frac{100 \text{ cm}}{0.5 \text{ cm}} = -200$$

So the angular magnification of the telescope is about –200, where the negative sign simply means that the image is inverted.

Chapter 10

Bouncing Light Waves: Reflection and Mirrors

*Y*ou can say a lot about mirrors — both plane (straight) mirrors and trickier spherical mirrors. As for spherical mirrors, you can get images in both *concave mirrors* (mirrors that look like the inside of a bowl) and *convex mirrors* (mirrors that look like the outside of the mirrored bowl). You can predict where images will form and whether they'll be upright or upside down — no mean feat, given that those mirrors can act pretty wacky when you hold them in your hand (or put them on a funhouse wall).

In this chapter, you work with some basic properties of reflection, and you see how light bounces off both flat and curved surfaces. After you see some ray diagrams, I present a couple of equations so you can get down with the math.

The Plane Truth: Reflecting on Mirror Basics

Even people with the most casual disregard for their appearance probably see themselves in a mirror every day. The flat plane mirror you use so frequently is also extremely important to optics. The basic law of how light reflects is expressed in terms of how light bounces off a plane mirror. Then, if you take any curved reflecting surface — like the ones in a carnival's hall of mirrors — and look at it closely enough, it appears flat at every point (just as the Earth is curved, but because you see it so close up, it appears flat wherever you're on it). So if you know how light reflects off a flat surface, you also know how light reflects off every part of any curved surface — bargain!

This idea applies wherever reflection occurs from a flat surface, even if it's not a mirror. So without further ado, here's your introduction to mirrors and other reflective surfaces.

Mirrors were often made of polished metal in the ancient world. These days, they're commonly made of metal electroplated onto glass. And glass itself can form a partial mirror — if you stand next to a window and look out, you often see a ghostly image of yourself looking out in the window glass. You see that image because glass commonly reflects about 7 percent of the light that hits it instead of transmitting it through the glass.

Getting the angles on plane mirrors

Figure 10-1 shows a *plane mirror* — that is, a straight mirror — lying on its back. A light ray comes in from upper left in the figure, hits the mirror, bounces off the mirror, and leaves to the upper right.

The angle at which the light ray comes in to the mirror is called the *angle of incidence,* θ_i, and the angle at which light is reflected is called the *angle of reflection,* θ_r. Note that these angles are with respect to the normal — a *normal* is a line perpendicular to the mirror's surface.

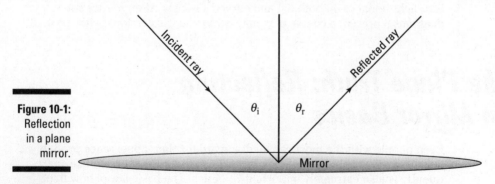

Figure 10-1:
Reflection
in a plane
mirror.

The angle of incidence is equal to the angle of reflection:

$$\theta_i = \theta_r$$

That's why when you're driving and you see an approaching car's image in your rear-view mirror, you know just which way to turn to see the actual car.

Forming images in plane mirrors

Plane mirrors are especially good at forming images. This section takes a look at image formation in a little more depth.

Mirrors form virtual images of objects, as you see in Figure 10-2. The image is *virtual* because no actual light rays meet to form that image (see Chapter 9 for more on virtual versus real images). In other words, you can't focus the image on a screen at the place the image appears to come from.

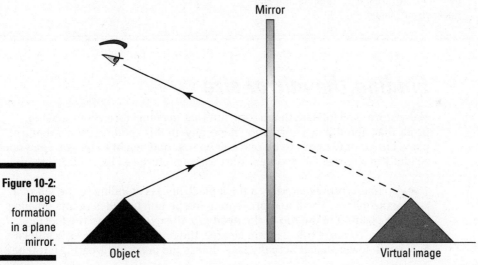

Mirror

Figure 10-2:
Image
formation
in a plane
mirror.

Object　　　　　　　　　　　　　　　Virtual image

Here's a set of observations you can make about an image formed in a plane mirror, besides the fact that it's virtual:

- The image is upright.
- The image is the same size as the object.
- The image is located as far behind the mirror as the object is in front of the mirror.
- The image is flipped back to front (see the nearby sidebar "Reversing a mirror myth: the left-right flip").

Reversing a mirror myth: The left-right flip

If you hold your right hand up to a mirror, you find that its image looks like a left hand (so you wouldn't be able to shake hands with this image — even if it were real and the mirror weren't in the way!). You may wonder why the image flips left and right without also flipping up and down. But it actually does neither of these. To see why, try the following experiment:

✔ Stand in front of a mirror and point your finger to the left; you see the image of your hand also pointing to the left, parallel to your pointing direction. Therefore, left and right are not flipped.

✔ Now point your finger straight up, and you see the image of your hand pointing straight up, parallel to your pointing finger. So up and down aren't flipped, either.

✔ Now try pointing away from yourself, straight into the mirror, and you see your image pointing straight toward you, out of the mirror — the complete opposite direction! So you can say that the mirror flips back to front.

Finding the mirror size

Many stores sell full-length mirrors, but that may be more about making profit than about image-formation necessity. In this section, I show that a plane mirror only needs to be one-half your actual height to let you see yourself fully in a mirror. To see this, start with a picture — Figure 10-3:

✔ A person (represented by a thick black line) is standing to the left of a plane mirror. The line representing the person includes points labeling the position of the top of the head *(T)*, the eyes *(E)*, and the feet *(F)*. (**Note:** To make the diagram clearer, the position of the eyes is shown much lower than it actually is — unless the person is wearing a top hat!)

✔ The vertical gray shaded line in the center represents a full-length plane mirror.

✔ The light rays leaving the person reflect from the mirror, creating an image, which is shown on the right as another thick, black line. The image has corresponding points labeled, showing the position of the image's top of the head *(T')*, eyes *(E')*, and feet *(F')*.

The points *A* and *B* show where on the mirror's surface the person sees the top of his or her head and feet, respectively. You can already tell that you don't need the whole length of this mirror to see all of yourself, because the distance *AB* is much less than the length of the mirror, *CD*. With a little geometry, you can work out exactly how big *AB* is — that is, how much of the full-length mirror you really need.

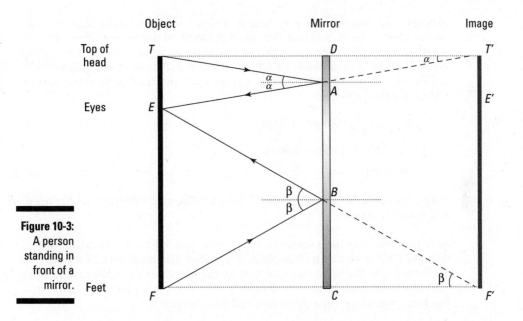

Figure 10-3:
A person
standing in
front of a
mirror.

Look again at Figure 10-3. You can see that the mirror is obeying the law of reflection: The angles of incidence and reflection from the rays from the top of your head, α, are equal. Then, because *T'E* is a straight line, the angle that it makes with your image must also be α. This means that the triangle *T'ET* is similar to triangle *T'AD* — because they're both right triangles that also share the angle α.

You also know that the image is as far behind the mirror as you are in front of it, so *T'A* is half the length of *T'E*. Because you have similar triangles, that means triangle *T'AD* is half the size of *T'ET* — which means that *AD* is half the length of *ET*:

$$AD = \frac{ET}{2}$$

You can do exactly the same thing for triangles *F'EF* and *F'BC*, because you can see that they're similar for the same reason. So *BC* is half the length of *EF*:

$$BC = \frac{EF}{2}$$

Now all you have to do to find the length of mirror you actually need is to subtract these two unused lengths *(AD* and *BC)* from the total *(CD)*:

$$AB = CD - AD - BC$$

Now try some numbers. If your height *(TF)* is 1.66 meters, how big does the mirror have to be so you can see your full length in the mirror? The full-length mirror, *CD,* is likewise 1.66 meters. Suppose your eyes are 0.06 meters below the top of your head *(ET)*. That means the distance from eyes to feet is then *TF – ET* = 1.60 meters. You can now work out the length of the part of the mirror that you use:

$$AB = 1.66 \text{ m} - \frac{0.06 \text{ m}}{2} - \frac{1.60 \text{ m}}{2}$$
$$= 1.66 \text{ m} - 0.03 \text{ m} - 0.80 \text{ m}$$
$$= 0.83 \text{ m}$$

This is half your height — you don't need a full-length mirror; you only need a half-length mirror.

Note that you may have seen this much more quickly by noticing that the triangle *T'F'E* is similar to triangle *ABE* and that *ABE* must be half the size of *T'F'E* (because the image is as far behind the mirror as the object is in front). Therefore, *AB* must be half the length of *T'F'*. Then, because your image is the same size as you are, *AB* is just half your height!

Working with Spherical Mirrors

A plane mirror makes an image that's the same size as the original object, at a position that's as far behind the mirror as the object is in front. When the mirror is curved, then the position, size, and orientation of the image can be very different. The inside or outside surface of a sphere creates such images. This is a convenient shape of mirror to study because it's simple, though you usually use only part of the surface of a sphere rather than the whole.

There are only two ways of looking at spherical mirrors — as convex and concave. Remember, if you're looking into a mirrored "cave," that's a *concave* mirror. Otherwise, it's *convex.*

Like lenses (see Chapter 9), spherical mirrors have a *center of curvature.* That's the center of the sphere that the mirror was cut from, and it's marked *C* in Figure 10-4. The distance from the center of curvature, *C,* to the mirror is called the *radius of curvature, R.* There's also a focal point, marked *F.* The *focal point* is where light rays that come in horizontally from the left end up being focused. The focal length is half of the radius of curvature, or looked at another way, *R* = 2*f*.

How do you handle spherical mirrors? You can draw ray diagrams that trace how several light rays travel from an object, bounce off the mirror, and end up forming an image (just as for lenses, which I cover in Chapter 9). In this section, I show you how to draw ray diagrams for both concave and convex mirrors.

Finding practical uses for curved mirrors

Spherical mirrors are used in many everyday devices, such as magnifying makeup mirrors and the security mirrors in shops. They're also used to turn the light from a light bulb into a beam in car headlights and flashlights. There's even a legend that Archimedes (the famous Greek mathematician who ran down the street shouting "Eureka!") had an idea to use curved mirrors as a weapon of war, focusing the sun's rays onto enemy ships and setting them on fire!

Curved mirrors are also used to make the largest telescopes in the world. Why mirrors? Because it's easier to build a large mirror than a large lens. Not only do you have to shape only

one side, but also you can support the large mirror all along its unsilvered side to stop it from curving further under its own weight.

Mirrors in telescopes aren't quite spherical. Anyone who has looked at his or her reflection in the back of a spoon or visited a hall of mirrors knows how the curves of a mirror can create very distorted images. When objects are very far from a spherical mirror, then the distortion is very small, but for the very fine level of precision required for astronomy, these distortions are too large, and corrections to the spherical curve are made to improve the image.

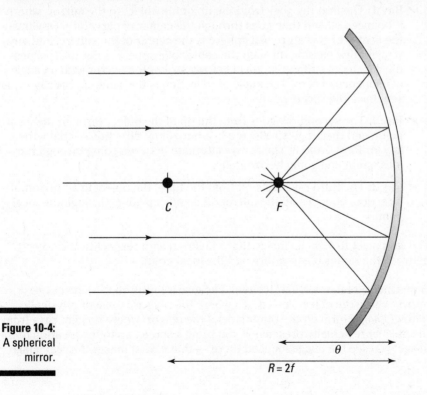

Figure 10-4:
A spherical
mirror.

Getting the inside scoop on concave mirrors

For a concave mirror, the part of the mirror that does the reflecting is on the inside of the spherical mirror. For concave mirrors, three different cases yield different types of images:

✔ The object is out farther than the center of curvature.

✔ The object is between the center of curvature and the focal point.

✔ The object is located between the focal point and the mirror itself.

This section takes a look at the various possibilities, starting by placing the object beyond the center of curvature and finding where the image forms.

Object farther out than the center of curvature

Figure 10-5 shows an object (represented by the thick arrow) being reflected in a concave mirror. Look at the ray diagram to see where the image will form in this situation and whether it's upright or upside down. Here's how the rays work:

✔ **Ray 1:** The first ray goes from the tip of the object to the mirror, where it bounces off and then goes through the center of curvature. Obviously, the center of curvature of a sphere is the center of the sphere, and any straight line passing through the center of a sphere is normal (perpendicular) to its surface, so this light ray strikes the mirror with an angle of incidence of zero. The angle of reflection is the same, so the ray is just sent back the way it came.

✔ **Ray 2:** The second ray goes from the tip of the object through the focal point, and then it gets reflected in a horizontal direction — that's the key for Rays 2 and 3: These rays alternate between going through the focal point and going horizontally.

✔ **Ray 3:** The third ray starts off from the tip of the object in a horizontal direction, bounces off the mirror, and ends up going through the focal point.

The rays meet to form an image that's inverted with respect to the object, between the radius of curvature and the focal point.

Is this image real or virtual? It's real, because it forms on the side of the mirror where the object is — that's where the rays are present physically (*virtual images* form on the other side of the mirror, where no light rays from the object are actually present). If you bring a screen up to the location of the image, you see the image focused there — that's what makes it a real image.

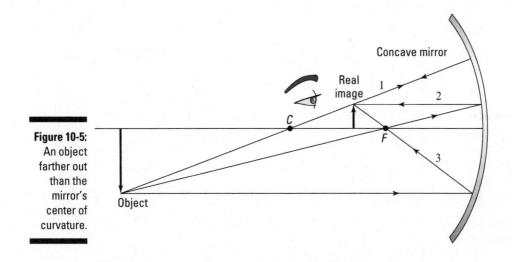

Figure 10-5:
An object
farther out
than the
mirror's
center of
curvature.

Object between the center of curvature and the focal point

Figure 10-6 shows an object being reflected in a concave mirror when the object is placed between the center of curvature and the focal point. Here's how to draw the three rays in the figure:

 ✔ **Ray 1:** The first ray goes from the tip of the object through the center of curvature to the mirror, where it's reflected back on its same path.

 ✔ **Ray 2:** The second ray travels from the tip of the object horizontally until it hits the mirror. Then it's reflected — and as is usual for rays that hit the mirror horizontally, it gets reflected through the focal point.

 ✔ **Ray 3:** The third ray travels from the tip of the object through the focal point, then to the mirror. When it's reflected from the mirror, the ray is traveling horizontally.

What's the net result? As you can see in Figure 10-6, the image is real (on the same side of the mirror as the object), inverted with respect to the object, and out past the center of curvature.

Object between the focal point and the mirror

Now for something really different — you end up with a virtual image in this case. If you place an object between the focal point of a spherical mirror and the mirror itself, all the rules change, because rays from the object can't pass through the focal point and then bounce off the mirror anymore.

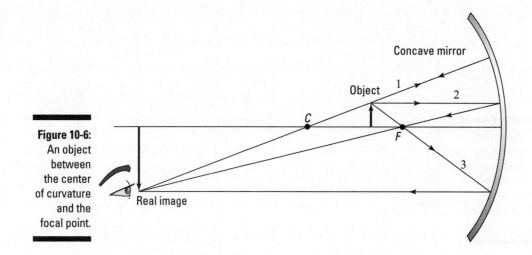

Figure 10-6:
An object between the center of curvature and the focal point.

As you can see in Figure 10-7, you're dealing with three rays:

- **Ray 1:** The first ray goes from the tip of the object to the center of curvature, and you extend the ray back to the mirror to complete this ray.

- **Ray 2:** The second ray goes from the tip of the object horizontally to the mirror — then it reflects from the mirror and goes through the focal point.

- **Ray 3:** The third ray is the tricky one. Normally, this ray goes from the tip of the object through the focal point and ends up going horizontally, but that's not going to work here, because if you send this ray through the focal point, it'll never hit the mirror. Instead, you send this ray from the tip of the object to the mirror *as though it were coming from the focal point*, as you can see in Figure 10-7. That does the trick.

Where do these rays come together? That's a trick question, because they don't come together at all — you have to extend the reflected rays behind the mirror itself. And with mirrors, that's the mark of a *virtual image* (that is, no light rays from the object penetrate behind the mirror, so the image that forms there is not actually caused by light rays that meet there — you can't bring a screen there and focus the image). So the image is virtual — and upright and magnified — as you see in the figure.

So the next time you're creating a salad in a mirrored metal bowl, take a look at what happens when you bring the lettuce close to the metal. As you pass the focal point, the image of the lettuce suddenly snaps into focus as upright and enlarged — and you get an image of some mega-sized lettuce.

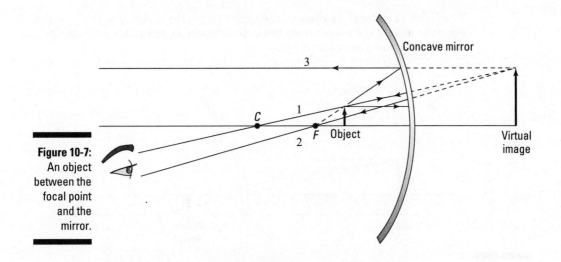

Figure 10-7:
An object
between the
focal point
and the
mirror.

Smaller and smaller: Seeing convex mirrors at work

Turn a mirrored bowl over so you're looking at the bottom of the bowl, and you have a *convex mirror.* Instead of bending light toward you, convex mirrors bend light away.

So what happens if you bring an object near a convex mirror? You can see the answer in ray diagrams in Figure 10-8. This time, the focal point and the center of curvature are on the other side of the mirror, so there's no question of different placement here (such as placing the object between the focal point and the center of curvature, placing the object closer than the focal point to the mirror, and so on). You can place the object only as Figure 10-8 shows — on the other side from the focal point and center of curvature.

You have the same three rays as with concave mirrors (see the preceding sections), but using them takes some fancy footwork. Here goes:

 ✔ **Ray 1:** The first ray goes from the tip of the object toward the center of curvature — but note that the center of curvature is on the other side of the mirror this time. That means that this ray goes from the tip of the object to the mirror and then bounces off in a way that it would as if it were coming from the center of curvature.

 ✔ **Ray 2:** The second ray travels from the tip of the object horizontally and then bounces off the mirror in a way that makes it appear this ray is coming from the focal point, as you can see in the figure.

 ✔ **Ray 3:** The third ray goes toward the focal point, which is on the other side of the mirror, and then bounces off the mirror, ending up going horizontally.

The result of all this? As you can see in the figure, the image is virtual (on the opposite side of the mirror from the object, where no light rays go), upright, and smaller than the object. Cool.

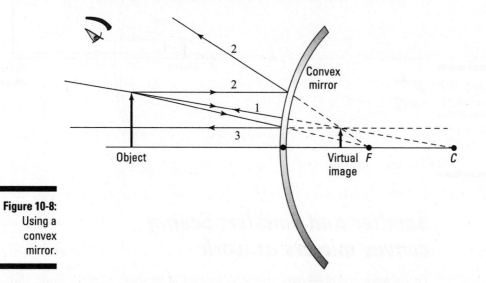

Figure 10-8:
Using a
convex
mirror.

The Numbers Roundup: Using Equations for Spherical Mirrors

Just as with the lenses (see Chapter 9), you can work out the location, size, and orientation of an image made by a spherical mirror with a couple of simple equations. These equations derive only from the *law of reflection* (the angle of incidence is equal to the angle of reflection) applied to every point of the curved surface of the mirror. But you don't have to worry about that here — I just give you the equations and show you how they work.

In this section, the mirror equation shows how the distances from the curved mirror to the object (d_o) and the distance from the mirror to the image (d_i) relate to the mirror's focal length (f). I also show you how to find magnification (m) when you know both d_o and d_i.

Getting numerical with the mirror equation

An equation for mirrors, cleverly called the *mirror equation,* relates the distance from the object to the mirror (d_o) and the distance from the mirror to the image (d_i) to the mirror's focal length (f). Here it is:

$$\frac{1}{d_o} + \frac{1}{d_i} = \frac{1}{f}$$

You may notice that, yep, this looks a lot like the thin-lens equation from Chapter 9. However, you have two ideas that are different from the thin-lens equation to keep in mind:

- ✓ The distance to the image, d_i, is negative if the image is on the other side of the mirror from the object. That is, d_i is negative if the image is virtual.

- ✓ The focal length, f, for convex mirrors is negative (that's just like the rule that the focal length for diverging lenses is negative, as I explain in Chapter 9).

The sign rules for mirrors are essentially the same as those for lenses: If the image is on the outgoing side (the side into which the rays are reflected by the mirror), the image distance is positive; otherwise, it's negative — just as for the lens. The sign rule for the focal length is reversed for mirrors because a convex mirror diverges parallel light rays — as does a concave lens — and vice versa.

d_i for an object between the focal length and the center of curvature

Try some numbers. Say you have a concave mirror that has a focal length of 5.0 centimeters, and you place an object 8.0 centimeters in front of it. Where does the image form? Start with the mirror equation:

$$\frac{1}{d_o} + \frac{1}{d_i} = \frac{1}{f}$$

Solve for the image distance, d_i, by rearranging the equation, combining the fractions, and simplifying:

$$\frac{1}{f} - \frac{1}{d_o} = \frac{1}{d_i}$$

$$d_i = \frac{1}{\frac{1}{f} - \frac{1}{d_o}}$$

$$d_i = \frac{1}{\frac{d_o - f}{fd_o}}$$

$$d_i = \frac{fd_o}{d_o - f}$$

Plugging in the numbers gives you the answer:

$$d_i = \frac{(5.0 \text{ cm})(8.0 \text{ cm})}{8.0 \text{ cm} - 5.0 \text{ cm}} \approx 13 \text{ cm}$$

So the result is positive, which means the image is real.

d_i for an object between the mirror and the focal length

How about this one? Say you have a concave mirror with a focal length of 5.0 centimeters, and you place an object 3.0 centimeters in front of it. Where does the image form? Use the mirror equation solved for the distance to the image (from the preceding section) like this:

$$d_i = \frac{fd_o}{d_o - f}$$

Plugging in the numbers gives you the following:

$$d_i = \frac{(5.0 \text{ cm})(3.0 \text{ cm})}{3.0 \text{ cm} - 5.0 \text{ cm}} = -7.5 \text{ cm}$$

So in this case, the distance to the image is negative — which means that image is virtual. That's as expected, because in this case, you're placing an object between a concave mirror and its focal point.

d_i for a convex mirror

Here's one more. Say this time that you have a convex mirror (not concave) with a focal length of –5.0 centimeters, and you place an object 7.0 centimeters in front of it — where does the image appear?

Note that in this case, the focal length is negative, which it must be for a convex mirror. You can use mirror equation solved for the image distance, d_i, like this:

$$d_i = \frac{fd_o}{d_o - f}$$

Putting in the numbers gives you

$$d_i = \frac{(-5.0 \text{ cm})(7.0 \text{ cm})}{7.0 \text{ cm} - (-5.0 \text{ cm})} \approx -2.9 \text{ cm}$$

So in this case, the image is virtual.

Discovering whether it's bigger or smaller: Magnification

The magnification equation gives you the amount an image is magnified with respect to the object — that is, the ratio of the image height over the object height. This equation for mirrors is just the same as it is for lenses, which I cover in Chapter 9:

$$m = -\frac{d_i}{d_o}$$

where d_i is the distance to the image and d_o the distance to the object.

Doing a magnification example

Suppose an object is 7.0 centimeters from a convex mirror (d_o = 7.0 centimeters). The mirror has a focal length of –5.0 centimeters, and your calculations (from the preceding section) tell you that d_i = –2.9 centimeters. What's the magnification of the image compared to the object? Use the magnification equation:

$$m = -\frac{d_i}{d_o}$$

Putting in the numbers gives you the answer:

$$m = -\frac{(-2.9 \text{ cm})}{7.0 \text{ cm}} \approx 0.41$$

In this case, the magnification is 0.41 — the image height is 0.41 times the object height, and the image is upright (because the magnification is positive).

Remember: Magnification doesn't always mean that the image is larger — the image here is smaller than the object (less than half the size).

Using the mirror equation and magnification equation together

In general, you have to use the mirror equation to find d_o first; then you can plug that into the magnification equation to find the magnification.

Say that you have a concave mirror, with a focal length of 8.0 centimeters, and you place an object 10.0 centimeters in front of it. What's the magnification of the image? First, you have to find the distance to the image. Use the equation mirror equation, solved for d_i:

$$d_i = \frac{fd_o}{d_o - f}$$

In this example, you get

$$d_i = \frac{(8.0 \text{ cm})(10.0 \text{ cm})}{10.0 \text{ cm} - 8.0 \text{ cm}} = 40 \text{ cm}$$

So d_i = 40 centimeters — it's positive, so the image is real. That means that the magnification is

$$m = -\frac{d_i}{d_o} = -\frac{40 \text{ cm}}{10.0 \text{ cm}} = -4.0$$

So the magnification is –4.0 — that is, the image is four times the height of the object. The magnification is negative, which tells you that the image is inverted.

So now you know everything about what's happening in this example — where the image appears, whether it's upright or inverted, whether it's real or virtual, and how big *i* is. Not bad for two little equations (the mirror equation and the magnification equation).

Chapter 11

Shedding Light on Light Wave Interference and Diffraction

. . . .

In This Chapter

▶ Understanding light wave interference

▶ Getting coherent light sources

▶ Looking at diffraction from a single slit

▶ Working with the multiple slits of diffraction gratings

▶ Finding resolving power when light passes through a hole

. . . .

*T*his chapter is all about *interference* — that is, what happens when light waves collide. Interference is a property of all waves (as you discover in Chapter 6). Because light waves are actually electromagnetic waves, their electric and magnetic fields can add or subtract if they overlap with each other. The resulting electric fields and magnetic fields can be stronger — or weaker — than the fields of either light wave alone, giving rise to some phenomena you may not expect. This chapter starts with the interference of two light waves and goes on from there.

Interference is the interaction of waves from a few sources, but the interference of waves from a great many sources is called *diffraction*. I discuss diffraction of sound waves in Chapter 7, where sound waves bend and spread when they fall upon a gap in a wall. In this chapter, you become much more familiar with diffraction for light waves, using Huygens's principle. This is a new way of thinking about how waves propagate, and it explains why they can bend around corners and spread out when they pass through gaps in walls. Huygens's principle shows how diffraction is really just the interference of a lot of waves and not something completely new.

When Waves Collide: Introducing Light Interference

When two or more light waves interfere with each other, it's called just that — *interference*. Interference occurs when the electric and magnetic fields of two or more light waves interact. The electric and magnetic fields of the two waves add together to give you a new wave.

To see interference, you have to have two light sources, one for each wave, that emit light of exactly the same wavelength. If you don't, then the relative phase between the light waves will change with time, and you won't see constructive interference (where the fields are in the same direction, resulting in an even stronger field) or destructive interference (where the fields are in the opposite direction and cancel each other out).

When two light sources emit the same wavelength of light continuously, they're called *coherent sources*. This is the importance of coherent light — it makes the interference effects of light observable:

✔ To get constructive interference, you need the peaks of the waves to line up.

✔ To get destructive interference, you need the peak of one wave to line up with the trough of another.

But generally, to get interference, you just need there to be a constant relation between the two phases. This can only come from coherent waves. In this section, you see how constructive and destructive interference work when you have two coherent light sources.

Meeting at the bars: In phase with constructive interference

Because light is indeed an electromagnetic wave, you know that it's made up of electric and magnetic fields. Figure 11-1 presents the electric fields for two light waves. Note what happens when they both end up at the same place, which I call point *P*, at the same time. As you can see, the two waves I'm adding are *in phase*. That means that when they meet, the peaks of one wave add to the peaks of the other, and the troughs of one wave add to the troughs of the other. That is, when two waves are in phase, they meet peak-to-peak and trough-to-trough.

The electric fields of the light waves simply add. The two waves have equal amplitude, so the resulting wave's peaks are twice as high, and its troughs are twice as low.

Getting ghostly images

Interference is at the heart of the workings of CDs, DVDs, and nowadays, Blu-ray Discs. Holographic images use interference to make three-dimensional images. But interference isn't always useful; sometimes it's an obstacle to be overcome.

Back in the day (before TV went digital), people who lived in metro areas with plenty of buildings often experienced this superposition of waves for themselves if they happened to have a television that had an antenna. The TV signals coming from the station would leave the station's antenna and come to the TV — but waves could also leave the station, bounce off a building, and come to the antenna. That meant the antenna had to deal with two signals — and they added up (or sometimes canceled). The TV

signal would appear to go all wacky, and people would get ghost images and shadows and the like — until they moved the antenna to get rid of the problem. Today's digital signal is carried by the same kinds of waves, so there's still interference, but it's not apparent in the same way. Digital signals are processed, so you lose the ghost images.

Interference is also an issue for phone networks. With so many mobile phones around, each sending and receiving radio signals, it'd be very easy for all the signals to interfere with each other, making one muddle of a signal. Some very clever technology overcomes this interference and makes sure all these signals stay separate.

When you add two light waves from coherent sources, their electric and magnetic fields add linearly. This is called the *principle of linear superposition.*

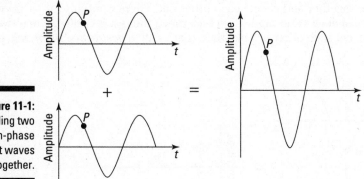

Figure 11-1:
Adding two
in-phase
light waves
together.

When two light waves collide as they do in Figure 11-1, peak-to-peak and trough-to-trough, and each light wave has the same wavelength (so they keep adding peak-to-peak and trough-to-trough as time goes on), the result is larger than either of the two waves by themselves; this process is called *constructive interference.* With constructive interference, the resulting light wave is stronger than either of its two components.

So if the two light waves that add in Figure 11-1 are in phase, the magnitude of the electric field from Wave 1 at point P is

$$E_1 = E_o \sin(\omega t)$$

And the magnitude of the electric field of light Wave 2 at point P is

$$E_2 = E_o \sin(\omega t)$$

That is, they reach their peaks and troughs at the same time, so they're in phase. When both these waves are present, then the total electric field is given by

$$E_1 + E_2 = 2E_0 \sin(\omega t)$$

This is just a wave which has the same frequency but twice the amplitude. This sum of the two waves is the *linear superposition*.

Going dark: Out of phase with destructive interference

Two waves don't need to meet peak-to-peak. They can meet out-of-phase, as Figure 11-2 shows. The waves are meeting at some point P, peak-to-trough and trough-to-peak. In other words, just when one wave is at its highest, the one it's interfering with is at its lowest, and vice versa.

In particular, the two waves in Figure 11-2 are as out of phase as they can be. They add together and cancel each other out. There's nothing left — they're opposites of each other, and when they meet, the result is zero. When two waves cancel each other out like that, that's called *destructive interference*.

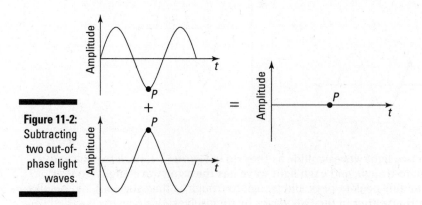

Figure 11-2: Subtracting two out-of-phase light waves.

Does this come in a medium?

In the 19th century, people thought that light, being a wave, must be carried by a medium. They believed that just as sound waves were carried by air, light waves were carried by a medium called *luminiferous ether*.

In 1887, Albert Michelson and Edward Morley used the following setup — an *interferometer* — and got a result that puzzled physicists. Michelson and Morley used this experiment to measure the movement of the Earth through the ether. They shone a beam of coherent light toward a semi-silvered mirror *(S)* that let half of the light pass straight through to the mirror at point M_2 while half reflected up toward the mirror at point M_1. The two parts of this separated beam were reflected back from these mirrors and recombined, having traveled in the two directions at 90° to each other. The mirror at M_1 could be moved so that these beams destructively or constructively interfered — that is, the path-length difference, $2(l_1 - l_2)$, was a whole number of wavelengths (constructive interference) or that plus half a wavelength (destructive interference). The light waves should travel with constant speed with respect to the ether.

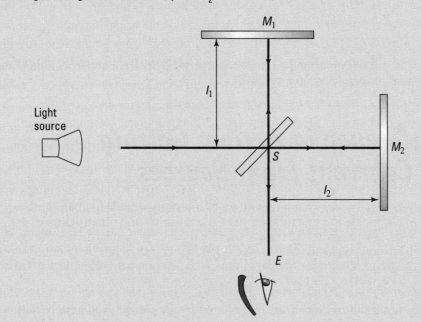

Suppose the setup was arranged so that the beams destructively interfered. The experimenters expected that the speed of the light in each direction would depend on the motion of the Earth through the ether. This would mean that when the mirror was rotated 90°, there should be a difference in the relative phases and they should no longer have destructive interference. But Michelson and Morley saw no difference whatsoever! This could only mean that there was no ether. Only when Einstein developed the special theory of relativity (which I cover in Chapter 12) did this result make sense.

In contrast to the waves in Figure 11-1, the two light waves in Figure 11-2 are as out of phase as possible. So if the first one looks like this at point *P*:

$$E_1 = E_o \sin(\omega t)$$

then the magnitude of the electric field of Wave 2 at point *P* is

$$E_2 = E_o \sin(\omega t + \pi)$$

In these equations, when one wave is hitting its peak, the other is hitting its trough. They're out of phase by an angle of π — and that's as out of phase as you can get.

When both these waves are present, then the total wave that results is just the sum:

$$
\begin{aligned}
&E_1 + E_2 \\
&= E_0 \sin(\omega t) + E_0 \sin(\omega t + \pi) \\
&= E_0 \sin(\omega t) - E_0 \sin(\omega t) \\
&= 0
\end{aligned}
$$

So you see that the two waves cancel out. The linear superposition of the two waves results in no wave at all.

Interference in Action: Getting Two Coherent Light Sources

Generally, when a number of waves interfere with each other, there are places where constructive interference occurs and other places where destructive interference occurs (see the earlier section "When Waves Collide: Introducing Light Interference" for info on types of interference). The result is a pattern of bright and dark areas, with intermediate brightness in between. This makes a pattern called the *interference pattern*.

Generally, to be able to see the interference pattern, you need to have a constant phase difference between the waves. For this, you need *coherent* light sources, which give you light waves that are of the same frequency. So how do you get two coherent light sources in the first place? In this section,

I discuss two methods: sending light from a single source through two slits or sending light through a thin film, using principles of reflection and refraction to split the light for you. I also show you the arrangement of the light and dark areas of the interference pattern in these cases.

Splitting light with double slits

Before the invention of lasers, one clever way to get two coherent light sources was to use the *same* source for both light rays by sending light through a double-slit arrangement.

If you send light of a particular color (and therefore of a particular wavelength) through the double slits, the two slits then act as two coherent light sources — each with the same wavelength. This kind of setup is called *Young's double-slit experiment* (credited to Thomas Young), and it provided some early proof of the wave nature of light. In this section, you see how double slits produce an interference pattern, predict where constructive and destructive interference occur, and look at some numbers.

Getting an interference pattern

When you send light from a single source through two slits, you now have two coherent light sources (the two slits, as Figure 11-3 shows). Those slits are arranged so that light from them falls on a screen, which you see at the right of the figure.

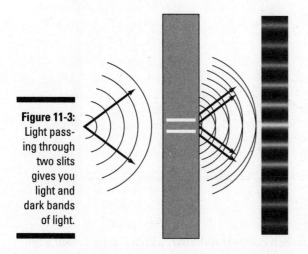

Figure 11-3: Light passing through two slits gives you light and dark bands of light.

Say that light from one slit illuminates a specific spot on the screen. Then light from the other slit falls on the same spot on the screen, and the two light waves interfere. Do they interfere constructively or destructively? That depends on how far the spot is from each of the two slits:

✔ **Constructive:** If the spot being illuminated is an integral number of light wavelengths from one slit, $m\lambda$, and it's $n\lambda$ from the other slit (where m and n are integers and n can equal m), then the two light waves hit the spot in phase — peak-to-peak, trough to trough. You end up with constructive interference, which causes a bright spot on the screen.

✔ **Destructive:** If the spot is at a distance from one slit that is an integral number of wavelengths, $m\lambda$, and the distance to that spot from the other slit is an integral number of wavelengths plus *one-half wavelength*, $(n + \frac{1}{2})\lambda$ from the other slit (where m and n are integers), then the two waves meet at the spot exactly out of phase, so there's destructive interference. The result is a dark spot.

You can see this situation more clearly in Figure 11-4. There, the distance between the slits is d, and the slits are a distance L from the screen. The curve at the screen represents the light intensity at every point (intensity is related to the mean of the squared electric field — see Chapter 8). The resulting light and dark regions, called the *interference pattern*, appears on the screen at right.

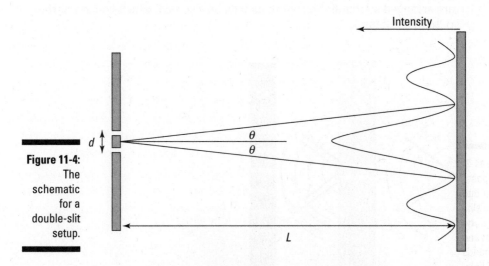

Figure 11-4:
The schematic for a double-slit setup.

The figure shows that you have a central bright bar, which is equidistant from the two slits, where constructive interference occurs. Then as θ increases, you have destructive interference between the rays from the two slits, and you get a dark bar. Then you get another light bar as the light rays end up in phase again — note that this light bar is less bright than the central bar.

That's what the interference pattern looks like — a central bright bar (also called a *fringe*) surrounded by dark bars and then alternating with successively diminishing light and dark bars. Here's how the naming of these bars works:

- ✔ The central bright bar is called the *zeroth-order bright bar* (or *zeroth-order bright fringe*).

- ✔ The next bright bar over is called the *first-order bright bar*. You have two of these, one on either side of the zeroth-order bright bar.

- ✔ The next bright bar is called the *second-order bright bar,* and so on.

Predicting where you get dark and light spots

Take a closer look at the double slits and the angle involved in Figure 11-5. To predict whether you end up with a light spot or a dark spot on the screen at a certain angle θ, you have to know the difference in how far that spot is from the two slits.

So what's the difference in the distance the light travels from each slit to the same spot on the screen? That difference in distance is marked as Δd in Figure 11-5. Because the screen is a long way from the slits, you can assume the two light rays are parallel, so each is emitted from their respective slits at the same angle θ. You also make the assumption that d is much less than L, the distance from the screen.

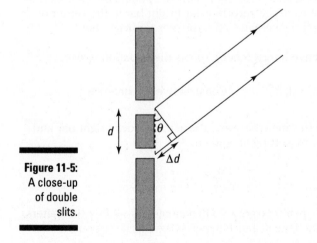

Figure 11-5: A close-up of double slits.

As you can see in the figure, the difference in distances from the two slits to the same spot of the screen is

$$\Delta d = d \sin \theta$$

So sin θ is equal to

$$\frac{\Delta d}{d} = \sin\theta$$

Note that when Δd is an integral number of wavelengths, you end up with constructive interference, which means that for all the bright bands in the interference pattern, the following equation holds true:

$$\sin\theta = m\frac{\lambda}{d} \qquad m = 0, 1, 2, 3, \ldots \qquad \text{Constructive interference}$$

And when Δd is an integral number of wavelengths plus one-half wavelength, you end up with destructive interference and a dark band on the screen. So for destructive interference, you have this relation:

$$\sin\theta = \left(m + \frac{1}{2}\right)\frac{\lambda}{d} \qquad m = 0, 1, 2, 3, \ldots \qquad \text{Destructive interference}$$

So now you have a handle on whether you get a bright bar or a dark bar at a certain angle from the two slits.

Trying some numbers for double slits

Say that you shine red light (λ = 713 nanometers) on two slits a distance of 2.00×10^{-4} meters apart, and an interference pattern appears on a screen 2.50 meters away. How far is it from the zeroth-order bright bar in the center of the interference pattern on the screen to the third-order bright bar?

For bright bars (constructive interference), this is the equation to use:

$$\sin\theta = m\frac{\lambda}{d} \qquad m = 0, 1, 2, 3, \ldots \qquad \text{Constructive interference}$$

In this case, you can find the angle between the zeroth-order bright bar and the third-order bright bar by setting m equal to 3:

$$\sin\theta = 3\frac{\lambda}{d}$$

Convert the wavelength to meters to get λ = 713 nanometers = 7.13×10^{-7} meters. Putting in the wavelength and the distance between the slits gives you

$$\sin\theta = (3)\frac{7.13 \times 10^{-7} \text{ m}}{2.00 \times 10^{-4} \text{ m}} \approx 1.07 \times 10^{-2}$$

Taking the inverse sine gives you θ:

$$\theta = \sin^{-1}(1.07 \times 10^{-2}) \approx 0.613°$$

That's a pretty small angle, but perhaps it'll come to something when you take into account how far away the screen is.

You know that the distance between the slits and the screen is 2.50 meters. That length forms the horizontal side of a right triangle where the vertical side is the distance between the bright bars that you're looking for and the angle between those two sides is θ. That means that if y is the length you're looking for, you have the following:

$$\tan\theta = \frac{y}{L}$$
$$y = L\tan\theta$$

Plugging in the numbers and doing the math gives you the answer:

$$y = (2.50 \text{ m}) \tan(0.613°) \approx 2.67 \times 10^{-2} \text{ m}$$

So the distance between the central bright bar and the third-order bright bar is 2.67 centimeters, which is roughly an inch. As you can see, even though red light has a very small wavelength, 7.13×10^{-7} meters, you still get a measureable effect when you position the screen far enough away from the double slits and position the double slits close enough together.

Gasoline-puddle rainbows: Splitting light with thin-film interference

Ever see some oily liquid like gasoline spilled on a puddle of water? If so, you probably saw rainbows of color form in the layer of gasoline. This same effect is responsible for the rainbows you see in soap bubbles. What you're really seeing is constructive and destructive interference patterns for different wavelengths of light — the constructive interference leads to a bright band of color. In this section, you take a look at how this process works — it's called *thin-film interference*.

Sending light rays on different paths

Suppose you have light going from air (where the index of refraction is $n_a = 1.00$) to gasoline ($n_y = 1.40$) and then to the underlying layer of water ($n_w = 1.33$), as Figure 11-6 shows (I discuss indexes of refraction in Chapter 9). At each step, there's some reflection, as you see in the figure — and the two rays heading off to the right end up interfering with each other, much as if they came from two coherent light sources.

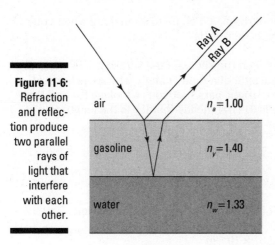

Figure 11-6:
Refraction
and reflec-
tion produce
two parallel
rays of
light that
interfere
with each
other.

Here's a blow-by-blow description of what happens with thin-film interference:

1. **Light travels through air to start, coming in from the upper left.**

2. **The light hits the air-gasoline boundary.**

   Some of the light is reflected from the boundary and heads off to the upper right.

   Most of the light continues into the gasoline and is refracted toward the normal (a perpendicular line) to the gasoline-air boundary.

3. **The light reaches the gasoline-water boundary, and some of the light bounces off the gasoline-water boundary and is reflected.**

4. **The reflected light hits the gasoline-air boundary.**

   Some light goes through the gasoline-air boundary and is refracted away from the normal. It ends up parallel to but horizontally displaced from the other light ray going off to the right.

5. **The two light rays that end up going off to the right interfere with each other, and the difference in the lengths of their paths (one ray goes through the gasoline) means that they can be out of phase.**

   When that path-length difference for the two waves is a multiple of the light's wavelength, you get constructive interference, and hence that region of the film is bright.

With thin-film interference, you assume that the thin film has a uniform thickness, but in practice, this isn't true — the thickness of a thin film of gasoline or soapy water in a soap bubble actually varies somewhat from place to place, which is why you get bands of color instead of one uniform color.

Accounting for changes in the wave's phase

When working with thin-film interference, you have to take one more effect into account besides the difference in path length: a phase change in the wave.

If you tie a rope to a wall and whip one end of the rope up and down, a pulse travels along the string to the wall. When it hits the wall, the rope reflects the pulse back — but first, the pulse is *inverted*. That is, it suffers a phase change of exactly one-half wavelength. So the pulse travels to the wall, hits the wall, gets inverted, and travels back to you, assuming there's some tension in the rope. On the other hand, if the end of the rope is just hanging free, a pulse is still at least partially reflected from the end of the rope, but there's no phase change.

Something similar happens to light at boundaries between materials that have different indexes of refraction:

- **Low to high:** When light, traveling through a medium, reflects from an interface with a material of higher refractive index (such as light in air reflecting from gasonline), there's a phase change in the reflected light of half a wavelength — half of the wavelength that the light would have in the material with the *higher* index of refraction.

- **High to low:** When light, traveling through a medium, reflects from an interface with a material of lower refractive index (such as light in gasoline reflecting from water), there's no phase change in the reflected light.

When this phase change occurs, you have to take it into account — it's as if the light ray traveled an additional one-half wavelength. Here's how you show that mathematically:

$$\text{Phase change} = \frac{\lambda_{\text{film}}}{2}$$

Doing some thin-film interference calculations

Say that there you are, filling up your car with gasoline, and notice that the previous customer was a little careless — some gasoline landed on a puddle next to your car. On closer inspection, you see that the film looks yellowish, and the time is just about noon, so the sunlight is hitting the gasoline film just about vertically. What's happening, and what minimum thicknesses of gasoline film on the water gives you this result?

Sunlight is pretty white because it's made up of all the wavelengths you normally see, from red to violet. The eye perceives white light with most of the blue part removed as yellow light, so if you see reflected sunlight from the gasoline film as yellowish, a lot of blue must be missing. (This is why the white sun ends up looking yellow — most of the blue wavelengths have been scattered away to make the sky blue!)

In other words, the gasoline film is just thick enough to give you destructive interference of blue light (which has a wavelength in air of 469 nonmeters). Wonderful, you're on your way to solving this problem.

What conditions give you destructive interference of the blue light? Well that's simple — if the light ray labeled A in Figure 11-6 is out of phase with Ray B, then they destructively interfere.

How does this phase difference come about? First, it happens because Ray B travels a greater distance than Ray A, because Ray B has to travel to the bottom of the gasoline film and reflect back up — a distance of about twice the thickness of the film. If the thickness of the film is t, then Ray B travels an extra distance of $2t$ compared to Ray A. As it travels this distance, Ray B makes a number of wave cycles that's equal to the number of wavelengths of the light in the distance traveled. For example, if the film is one wavelength thick, then Ray B goes through two cycles as it travels to the bottom and reflects back up (remember the extra distance covered is $2t$).

If Ray B goes through a whole number of cycles as it travels through the film of gasoline, then

$$2t = m\lambda_{gas} \qquad m = 1, 2, 3, \ldots$$

where λ_{gas} is the wavelength of the blue light in the gasoline and m is a whole number.

If this is the case, then Ray A and Ray B are in phase and constructively interfere, right? Nope. You need to account for the phase shifts that can occur when light reflects. Ray B reflects from within the gasoline film off the gas-water surface (Step 3 in the preceding section). But because water has a lower refractive index than gas, there's no phase shift for this ray. However, Ray A reflects from the air off the air-gasoline surface (Step 2). Because the gasoline has a higher refractive index, this ray undergoes a phase shift of half a cycle. So in this case, Ray A and Ray B are now out of phase when the preceding equation holds, in which case the rays destructively interfere.

All you need to do now is work out the wavelength of the blue light in gasoline. The key is to realize that the frequency of the light is always the same, whatever the material it's traveling through. Only its speed and wavelength change for different refractive indexes.

If you take the equation for the refractive index of a material and divide the top and bottom of the fraction by the frequency, f, you get the following:

$$n = \frac{c}{v} = \frac{\dfrac{c}{f}}{\dfrac{v}{f}}$$

But because you know that the wavelength is just the speed divided by the frequency, you can write this as

$$n_{gas} = \frac{\lambda_{air}}{\lambda_{gas}}$$

Rearrange this and write an equation for the wavelength of the light in gasoline:

$$\lambda_{gas} = \frac{\lambda_{air}}{n_{gas}}$$

Put in the numbers to work out the wavelength of the blue light in gasoline. You know the wavelength of the blue light in air is λ_{air} = 469 nm and the refractive index of gasoline is n_{gas} = 1.40, so

$$\lambda_{gas} = \frac{469 \text{ nm}}{1.40} = 335 \text{ nm}$$

At last you can work out how thick the gasoline film needs to be to give you destructive interference for the blue light and so make the sunlight appear yellow. From earlier in this section, you know that the thickness has to be related to the wavelength of the blue light in gasoline by

$$2t = m\lambda_{gas} \qquad m = 1, 2, 3, \ldots$$

The minimum occurs when m = 1, in which case the thickness of the film is

$$t = \frac{335 \text{ nm}}{2} \approx 168 \text{ nm}$$

So there you have it — the gasoline film needs to be 168 nanometers thick so that you see yellow light in the puddle.

Single-Slit Diffraction: Getting Interference from Wavelets

People usually think of light traveling in straight lines. However, in certain circumstances, light can bend around corners to reach places it couldn't go if it traveled only in a straight line (just as sound waves can bend around corners; see Chapter 7). You don't usually notice this effect because the small wavelengths of light usually make this bending quite small.

This bending, called *diffraction,* comes from the interference of a very large number of waves. In this section, I explain why light spreads out when it passes through a single slit, and you see the strange patterns of light and dark created when it does. You also discover a way to use this effect to make precise measurements of wavelength.

Huygens's principle: Looking at how diffraction works with a single slit

Figure 11-7 shows a single slit and the intensity of light as it appears on a screen some distance away from the single slit. How on Earth do you get interference from a single slit? This process is called *diffraction,* and it depends on the idea that every point on the front of a wave acts like a coherent source of light. All those point sources that make up a wave front are responsible for the interference pattern.

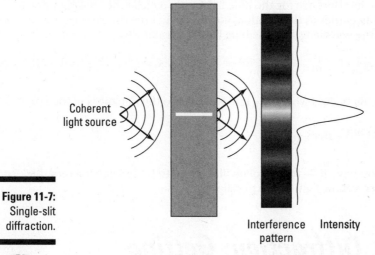

Figure 11-7:
Single-slit
diffraction.

Coherent
light source

Interference   Intensity
pattern

Diffraction all comes about because of *Huygens's principle*, which says the following:

> "Every point on a wave front acts as a source of wavelets that move forward with the speed of the overall wave; at any later time, the wave is that surface that is tangent to all the moving wavelets."

So every point of the wave front going through a single slit (of some width W), as you see in Figure 11-8, acts as a coherent source of wavelets. If that light then strikes a screen, the result of all the wavelets is a pattern of light like the one in Figure 11-7 — a wide central bright bar or fringe, flanked by successively smaller bright bars, with dark bars in between them. If light didn't obey Huygens's principle, then you'd get no pattern from a single slit — you'd just see the image of the single slit on the far screen.

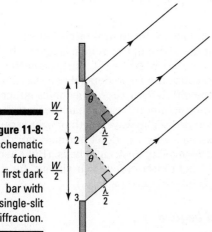

Figure 11-8:
A schematic for the first dark bar with single-slit diffraction.

Getting the bars in the diffraction pattern

So how do you get dark bars in the diffraction pattern when you have light going through a single slit? This section explains where all those dark bars come from.

Arriving at the first dark bar

Look back at Figure 11-8. There, light traveling from the top of the slit, Point 1, arrives at the screen exactly out of phase by one-half wavelength from light traveling from Point 2. That means that the light from Points 1 and 2 cancel each other out on the screen, interfering with each other destructively. (See the earlier section "Going dark: Out of phase with destructive interference" for the basics on destructive interference.)

In fact, every ray of light from the top of the slit is canceled out by a ray of light from the bottom half of the slit, which arrives exactly one-half wavelength out of phase — so you get the first dark bar in the diffraction pattern.

Think of the slit in Figure 11-8 in two equal sections. Consider the wave from Points 1 and 2. You get destructive interference between these two waves if they have a path-length difference of half a wavelength. The shaded right triangle shows that the path-length difference between these two waves is then given by

$$\frac{W}{2}\sin\theta = \frac{\lambda}{2}$$

Then the waves from every point between 1 and 2 destructively interfere with the wave from the corresponding point between 2 and 3. So you can write the angle of the first dark band as

$$\sin \theta = \frac{\lambda}{W}$$ First dark bar in the diffraction pattern

Most of the light passing through the slit falls in the central bright region, between the first dark bars. Notice that the width of this region is inversely proportional to the width of the slit — the narrower the slit, the greater the angle over which this light is spread. Because the angle depends on the ratio of wavelength to the width of the slit, the diffraction pattern becomes noticeable only when the width of the slit is not too much greater than the wavelength of the light (so that λ/W is not too tiny). The wavelength of light is quite small by everyday standards, which explains why you don't normally notice this effect and most people think of light traveling in straight lines. (But check out the nearby sidebar "Street smarts: A light interference experiment" for a really cool way you can actually see this interference pattern.)

Getting to the second dark bar and beyond

Take a look at the situation in Figure 11-9, where you have light passing through a single slit and you're creating the second dark bar in the diffraction pattern. Light rays from Point 1 are canceled by light rays from Point 2, which arrive at the screen exactly one-half wavelength out of phase. Light from Point 3 is canceled by light from Point 4, and so on.

Street smarts: A light interference experiment

You can make destructive interference happen very easily now that you know what to look for. Here's how: Wait till nighttime, and find yourself a distant streetlight. Then put your finger and thumb together to make a very small gap between them. Carefully look at the streetlight through this gap. With some very careful adjustments and a steady hand, you'll be able to see the interference pattern. Your finger and thumb make the slit, and your eye acts like the screen.

If you're really careful, you can even adjust the width of the gap between your finger and thumb and watch the width of the central peak grow as you make the gap smaller. I hope you try it — seeing this can be a little tricky, but you can do it if you have a steady hand.

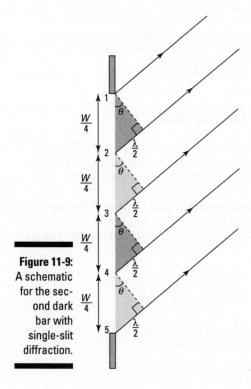

Figure 11-9:
A schematic for the second dark bar with single-slit diffraction.

Here, you can see the relation that connects W, λ, and θ from the triangle whose two short sides are W and 2λ. If you look at the shaded triangle between Points 1 and 2, then you have destructive interference if the path-length difference is given by

$$\frac{W}{4}\sin\theta = \frac{\lambda}{2}$$

This applies to any point and the corresponding point $W/4$ down from it. So then you can say that the angle of the second dark band is given by

$$\sin\theta = \frac{2\lambda}{W} \quad \text{Second dark bar in the diffraction pattern}$$

You can progress to higher-order dark bars as well, and you end up with this equation, which gives you the angle at which dark bars appear on the screen:

$$\sin\theta = \frac{m\lambda}{W} \quad m = 1, 2, 3, \dots \quad \begin{array}{l}\text{Dark bars in a single-slit} \\ \text{diffraction pattern}\end{array}$$

So there you have it — now if you know the width of a single slit, you can figure out where the dark bars will appear in the diffraction pattern. And because the central bright bar is straddled by first-order dark bars, you can figure out the width of the central bright bar as well.

Doing diffraction calculations

Say that you have a single slit width, $W = 5.0 \times 10^{-6}$ meters, and it's $L = 0.5$ meters away from a screen. You shine blue light ($\lambda = 469$ nanometers) on the single slit. What is the width of the central bright bar in the diffraction pattern?

The central bright bar is straddled by first-order dark bars, so if the distance to the first dark bar is y, then the width of the central bright bar is $2y$. So all you need to find is the distance to the first dark bar, and you can use the following equation for that:

$$\sin\theta = \frac{m\lambda}{W} \qquad m = 1, 2, 3, \dots \qquad \text{Dark bars in a single-slit diffraction pattern}$$

For the first dark bar, $m = 1$. A nanometer is a billionth of a meter, so 469 nanometers $= 4.69 \times 10^{-7}$ meters. So here's what you have for the first dark bar:

$$\sin\theta = \frac{\lambda}{W} = \frac{4.69 \times 10^{-7} \text{ m}}{5.0 \times 10^{-6} \text{ m}} \approx 0.094$$

Taking the inverse sine gives you the angle:

$$\theta = \sin^{-1}(0.094) \approx 5.4°$$

So that's the angle at which the first dark bar appears. You still need to find y, the distance of the first dark bar from the center of the diffraction pattern. Because the distance to the screen is L, you have the following equation:

$$\tan\theta = \frac{y}{L}$$

which you can rearrange like this:

$$y = L \tan\theta$$

Because $L = 0.50$ meters, you have

$$y = (0.50 \text{ m}) \tan 5.4° \approx 0.047 \text{ m}$$

Okay, so the first dark bar appears at 0.047 meters, or 4.7 centimeters, from the center of the diffraction pattern. You need to find $2y$ to get the width of the central bright bar, so multiply by 2 to get

$$2y = 2\ (4.7\ \text{cm}) = 9.4\ \text{cm}$$

So in this case, the central bright bar is 9.4 centimeters wide. Cool.

Multiple Slits: Taking It to the Limit with Diffraction Gratings

A *diffraction grating* has many slits — hundreds, thousands of slits. You're far beyond double slits now. A diffraction grating has so many slits that they're measured in slits per centimeter — and 40,000 slits per centimeter is not unusual for a diffraction grating. That's a lot of slits.

Diffraction gratings are frequently made from plates of glass or something equally transparent and are incised with a diamond-tipped, machine-controlled scribe. The scribe draws lines to the tune of 40,000 per centimeter. The slits are the clear places between the lines.

A diffraction grating works in the same way as single slits and double slits — through interference. Each slit acts as a coherent source of light. In this section, you see how diffraction gratings work and how physicists use them to separate colors.

Separating colors with diffraction gratings

Diffraction gratings are great for determining exactly which wavelength of light you're dealing with. When you have a single slit or double slit, the bright bars you get on the screen are pretty broad, making accurate measurements of the angle (and hence, the wavelength) difficult. If you're trying to find the exact center of a bright bar that's 4.0 centimeters in width, there's a lot of room for error.

Diffraction gratings are different, however. You end up with very sharp, very narrow bright bars, which are called *maxima* (plural of *maximum*) in diffraction-grating speak. The bright bars from a single slit are wide, those from a double slit are a little less wide, and the bright bars from a diffraction grating are razor thin.

Besides the main bright bars in the pattern generated by a diffraction grating, you also have some other, secondary bars due to the diffraction of light passing through each of the single slits. But although the secondary bright bars are significant when you have a double-slit setup, they're almost invisible when you're using a diffraction grating. All you see are the *principal maxima.*

To get a maximum in the diffraction grating pattern, light from the top slit travels from that slit to the screen. Light from the next slit down must travel that same distance plus one wavelength. Light from the next slit down must travel the same distance as from the top slit to the spot on screen, plus two wavelengths, and so on. When light travels a distance that's one wavelength longer than the path from the slit directly above it, you have constructive interference — that is, a maximum — at that spot.

By analogy with what I show you earlier for single and double slits in the sections "Splitting light with double slits" and "Single-Slit Diffraction: Getting Interference from Wavelets," you get the following relation for maxima in diffraction grating patterns:

$$\sin\theta = \frac{m\lambda}{d} \qquad m = 0, 1, 2, 3, \ldots \qquad \text{Maxima for diffraction gratings}$$

where d is the distance between the slits in the grating and θ is the angle from the center of the diffraction grating to the spot on the screen you're looking at. Looking at this relation, you can see that you have a central maximum ($m = 0$), another maximum right next to it ($m = 1$), and then other maxima ($m = 2, 3$, and so on).

Trying some diffraction-grating calculations

Say that you have a diffraction grating with 10,000 slits per centimeter and that you send a mixture of light through it, half violet light ($\theta = 410$ nanometers) and half red light ($\theta = 660$ nanometers). Try showing that the diffraction grating breaks down the light so that the two components, red and violet, are clearly separated.

To solve this problem, you can find the first-order maximum for each color of light, red and violet, and show that their angles vary significantly. For the first-order maxima, $m = 1$ in this relation:

$$\sin\theta = \frac{m\lambda}{d} \qquad m = 0, 1, 2, 3, \ldots \qquad \text{Maxima for diffraction gratings}$$

So if $m = 1$, you have

$$\sin\theta = \frac{\lambda}{d}$$

What's the angle for the first-order maximum? Taking the inverse sine gives you

$$\theta = \sin^{-1}\left(\frac{\lambda}{d}\right)$$

So when the diffraction grating has 10,000 slits per centimeter, that means that the distance between each slit is

$$d = \frac{1}{10,000 \text{ slits/cm}} = 1.0 \times 10^{-4} \text{ cm}$$

For violet light, you have the following (410 nm = 4.1×10^{-5} cm):

$$\frac{\lambda}{d} = \frac{4.1 \times 10^{-5} \text{ cm}}{1.0 \times 10^{-4} \text{ cm}} = 0.41$$

And taking the inverse sine of this gives you

$$\theta = \sin^{-1}\left(\frac{\lambda}{d}\right) = \sin^{-1}(0.41) \approx 24°$$

For red light, you have the following (660 nm = 6.6×10^{-5} cm):

$$\frac{\lambda}{d} = \frac{6.6 \times 10^{-5} \text{ cm}}{1.0 \times 10^{-4} \text{ cm}} = 0.66$$

And taking the inverse sine of this gives you

$$\theta = \sin^{-1}\left(\frac{\lambda}{d}\right) = \sin^{-1}(0.66) \approx 41°$$

So the first principal maximum from violet light is at about 24°, and the first principal maximum from red light is at about 41° — that's a very wide separation in angle, so you can tell quite clearly the makeup of the light you're studying.

Seeing Clearly: Resolving Power and Diffraction from a Hole

Here's an interesting point: Light traveling through the lens of a camera treats that lens much as it would a single slit, only in circular form, which means you end up with a single-slit diffraction pattern on the film — that is, the image appears a little blurry because of single-slit diffraction (see the earlier section "Single-Slit Diffraction: Getting Interference from Wavelets" for details on diffraction).

In Figure 11-10, light from two objects is passing through a circular aperture (much like a lens) and falling on a screen. So how small can you make the circular aperture and still be able to distinguish between the two objects on the screen? In other words, how small can you make the hole and still see the images of the two objects as separate? This value is known as the *resolving power* of a circular aperture.

The rule is that you're just about at the limit of resolving power for two objects when the first dark bar from one image overlaps the central bright bar of the second image, as Figure 11-10 illustrates.

Figure 11-10:
Resolving
power,
where an
image's first
dark bar
overlaps
another
image's
central
bright bar.

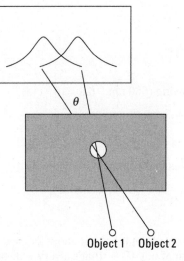

Object 1   Object 2

If the first dark bar of one image overlaps the central bright bar of the other image, doesn't that dim the bright bar? No, not at all. Keep in mind that a dark bar in a diffraction pattern doesn't mean a shadow or anything like that — it just means that no light comes from the corresponding source at that point. So when the first image's dark bar overlaps a bright bar, it just means that no light from the first image falls there.

When the first dark bar of one image overlaps the central bright bar of the other image, the angle between the light rays from the two objects turns out to be the following:

$$\sin\theta = 1.22\frac{\lambda}{D}$$

where θ is the angle (which represents the minimum angle between two objects such that you can still distinguish them), λ is the wavelength of the light, and D is the diameter of the aperture. So for a particular diameter of circular aperture and a particular wavelength, θ is the minimum angular separation between two objects such that you can still see them as distinct.

For example, if two objects are 100 meters away from you, how far apart must they be from each other so that you can still tell them apart? For this example, work with green light, which is the exact center of the visible spectrum, λ_{air} = 555 nanometers. The pupil of your eye is about 3.0 millimeters.

So are you ready to use the resolving-power equation? Not quite, because although you know the wavelength of light you're working with in air, you don't know the wavelength of that light where it counts — in the eye. To figure that out, use the equation derived earlier in "Doing some thin-film interference calculations":

$$\lambda_{eye} = \frac{\lambda_{air}}{n}$$

where λ_{eye} is the wavelength of the light in the eye. The index of refraction of the clear medium in the eye turns out to be pretty close to water — 1.36 (compared to 1.33 for water). So plugging in the numbers, you get the following wavelength:

$$\lambda_{eye} = \frac{\lambda_{air}}{n} = \frac{555 \text{ nm}}{1.36} \approx 408 \text{ nm}$$

Now you're ready to find the resolving power. Note that for small angles, sin θ = θ if you measure θ in radians, so you have the following equation for resolving power:

$$\theta \approx 1.22 \frac{\lambda}{D} \text{ radians}$$

The eye's pupil is 3.0 millimeters, or 3.0×10^6 nanometers. Plugging in the numbers gives you the answer:

$$\theta \approx 1.22 \frac{\lambda}{D} = 1.22 \frac{408 \text{ nm}}{3.0 \times 10^6 \text{ nm}} \approx 1.7 \times 10^{-4} \text{ radians}$$

So you can resolve (theoretically) an angle of 1.7×10^{-4} radians. If you're 100 meters from the objects, what does that work out to be in terms of distance?

With such a small angle, the distance the angle translates to is just the angle (in radians) multiplied by how far away you are, so you have the following:

$$(1.7 \times 10^{-4})(100 \text{ m}) = 1.7 \times 10^{-2} \text{ m}$$

So from 100 meters away, you can (theoretically) resolve two objects as distinct if they're at least 1.7 centimeters apart.

Part IV
Modern Physics

1886 – 7-year-old Albert Einstein's genius begins to reveal itself.

In this part . . .

This part covers exciting topics you may have been waiting for: special relativity (that is, Einstein's ideas about what happens near light speed), radioactivity, quantum physics, and matter waves. You see everything from the spectrum of hydrogen to the famous $E = mc^2$ equation.

Chapter 12

Heeding What Einstein Said: Special Relativity

*W*elcome to special relativity, the topic that made Albert Einstein famous. In this chapter, you get a handle on the strange but true facts of special relativity, where few things are as they seem. You see how frames of reference make certain values vary, depending on who's doing the measuring, and you look at what happens at speeds approaching the speed of light. Time slows, lengths contract, and mass can be converted into energy ($E = mc^2$ — you knew that was coming up in this book, didn't you?). You shouldn't worry if the results of this chapter seem strange to you — in fact, if you can appreciate how strange they are, then you're well on your way.

Special relativity gives you lots of weird and wonderful results. One thing, however, always seems to disappoint: the fact that the speed of light is the ultimate speed. Yep, it's true until proven otherwise. So if you're looking for true faster-than-light travel, sorry — you won't find it in special relativity. (I can sympathize with the letdown; I'm a *Star Trek* fan.)

However, discussions of relativity still leave plenty of room for imagination. Never mind that space shuttles fall far short of the speed of light or that you're lucky if anything traveling that fast registers as a blur. For these examples, you can ignore the limits of technology and the human body and pretend that you're working with a really great crew with really great gear, so the only elements at play are the principles of physics. So yes, you *can* see exactly what's going on in that fast, transparent rocket ship from where you're standing. Ready? All systems go.

Blasting Off with Relativity Basics

So what was special relativity created to deal with, and what makes it so special? Newton's laws of motion work fine for the speeds you experience in everyday life, but you need a new way to describe anything traveling near the upper limit of speed: the speed of light in a vacuum. That's where Einstein's theory of special relativity steps in.

In this section, you first explore what's *special* about this theory, what's *relative,* and what those ideas have to do with frames of reference. Then you look at two postulates on which Einstein based this theory. These postulates are simply assumptions from which the rest of the theory follows. They're both based on the results of precise experiments. One postulate is not surprising at all, but the other is a little bit odd, and in combination with the first, it meant physicists had to change their precious old ideas about the very nature of space and time.

Start from where you're standing: Understanding reference frames

Special relativity is a theory that predicts how events are measured with respect to various observers who can be in motion with respect to the event. An *event* is just a physical happening, such as the explosion of a firecracker or the ticking of a clock or a train's passing a certain point. Events happen at a particular place and time, as measured by the people who observe them. Those people can be moving with respect to the event, or they can be stationary with respect to the event.

For example, in Figure 12-1, a firecracker goes off, causing a flash of light. That's an event, and two observers watch it. One observer is stationary with respect to the firecracker, and the other observer is moving in a rocket — in a straight line at a constant speed — with respect to the firecracker. (Rockets figure a lot in discussions of special relativity.)

Each observer carries his or her own coordinate system, as the figure shows, and the observers each measure the event — its location and time — with respect to their individual coordinate systems and clocks. So each observer gets his or her own x, y, and z coordinates for the event and his or her own time, t. Special relativity is all about relating the measurements that the two observers make.

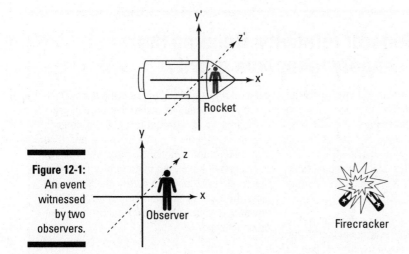

Figure 12-1:
An event
witnessed
by two
observers.

The coordinate systems and clocks that the observers carry with them are called their *reference frames,* or their *frames of reference.* All measurements an observer makes are with respect to his or her reference frame — and each observer is at rest with respect to the coordinates in the reference frame.

So why is it "special" relativity? Each observer's reference frame is a special type of reference frame — an inertial reference frame. An *inertial* reference frame is just one that's not accelerating. That is, Newton's law about how objects in motion tend to stay in motion and objects at rest tend to stay at rest holds in inertial reference frames.

Not all reference frames are inertial. For example, a reference frame that's rotating isn't inertial, because it's accelerating. Life in a noninertial reference frame may sound pretty surprising — perhaps you could put an object on the ground and then watch it suddenly start to drift away for no apparent reason, because your reference frame would be accelerating.

Technically, standing on the surface of the Earth does not put you into an inertial reference frame, because the Earth has gravity, so the whole reference frame is undergoing acceleration due to gravity. And of course, the Earth is spinning, so there's centripetal acceleration, and it's wobbling, so there's wobbling acceleration, and it's going around the sun, and so on. But for the sake of simplicity, you can ignore all those effects in this chapter and treat observers standing on the ground as being in their own reference frames, temporarily ignoring the force of gravity and so on.

General relativity: Bending the theory to include gravity

In special relativity, if a body is moving without an external force, then it'll continue to move in a straight line at constant speed — this is *inertial motion.* Anything undergoing inertial motion follows a straight line in an inertial frame. You may like to think of a ball rolling on a pool table (if you forget about friction); the ball rolls along in a straight line at constant speed.

Einstein made an astonishing extension to the theory of special relativity to account for gravity: *general relativity.* In this theory, inertial motion can undergo acceleration if a gravitational field is around. You may think that the force of gravity counts as an external force, but it doesn't. Gravity is actually a *curve* in space and time that makes inertial motions accelerate. Think of the pool ball again, but instead of a flat table, there's a curve in it, like a very shallow basin. Now the path of the ball isn't a straight line anymore — this is like inertial motion in a gravitational field.

Looking at special relativity's postulates

When Einstein came up with the theory of special relativity, he started with two *postulates,* or assumptions, that the theory rests on. The postulates are these:

- ✔ **The relativity postulate:** The laws of physics are the same in all inertial reference frames.

- ✔ **The speed of light postulate:** The speed of light in a vacuum, *c,* always has the same value in any inertial reference frame, no matter how fast the observer and the light source are moving with respect to each other.

In this section, I discuss the significance of both concepts.

The relativity postulate

The *relativity postulate* tells you that one inertial reference frame is as good as another, and you can't tell them apart with experiment. For example, if you're in one inertial reference frame and someone else is in another, you can't tell those frames apart through tests. So if you're in the rocket back in Figure 12-1, it's just as valid to say that the Earth is moving toward you as it is to stand on the Earth and say that rocket is moving toward you. That's what relativity is all about.

This also means that there's no "absolute reference frame" where objects are at "absolute rest." All that matters is the relative motion of reference frames.

The speed of light postulate

The *speed of light postulate* says that the speed of light in a vacuum, *c*, is always *c* — even if that light comes from an inertial reference frame that's moving toward you at half the speed of light.

Although the relativity postulate (which says that the laws of physics are the same in all inertial reference frames) isn't hard to swallow, the speed of light postulate is harder to accept. After all, if you're in one car going 5 meters per second and are approached by another going 10 meters per second, you're moving at 15 meters per second with respect to the other car. So if you're standing by the side of a road and a car approaches you at 10 meters per second with its headlights on, wouldn't you measure the speed of light from the headlights as *c* + 10 meters per second?

That's not how it works with the speed of light, however, or with speeds approaching the speed of light. As the postulate says, the speed of light from the headlights of the car coming toward you would be *c*, not *c* + 10 meters per second. This is an extraordinary result, and experiment has verified it over and over.

To come up with relativity, Einstein used James Clerk Maxwell's equations for electromagnetic waves, which predict that the speed of light in a vacuum is a constant given by

$$c = \frac{1}{\left(\mu_o \varepsilon_0\right)^{1/2}}$$

where μ_o and ε_0 are the permeability and permittivity of space (see Chapter 8). Einstein was the first to take this equation for what it means — that any observer will measure this constant value for the speed of light.

However, even though the speed of light is constant, you can have a shift in the frequency and wavelength of light from moving sources — for details, see the later sidebar "Red light, blue light: Shifting light frequencies."

Seeing Special Relativity at Work

To understand special relativity, you have to change the way you think about space and time. For instance, the length of time between two events depends on the observer (and not just because some people are in different time zones or have slow reflexes or faulty watches). You may think this bizarre. If you ask someone what time it is, isn't there only one right answer? Nope! The distance between two events also depends on the observer, which is just as strange but true.

In this section, you explore time dilation and length contraction. You also start to look at how all this affects Newtonian mechanics by asking about the momentum of a particle in relativity. Of course, these ideas become apparent only at very high speeds — nearing the speed of light. If humans had evolved to be able to walk at nearly the speed of light, or if the speed of light was much slower, then special relativity would already make sense to you.

Slowing time: Chilling out with time dilation

Say you're just standing around on the surface of the Earth, and a rocket passes by overhead at great speed. You're in touch with the people on the rocket, whom you've instructed to read off seconds from a clock as they pass you. When they read off the seconds, however, you notice that somehow, their seconds are longer than your seconds. What's going on? Time dilation.

Time dilation is the phenomenon predicted by the theory of special relativity that says time in two inertial reference frames moving with respect to each other will appear to be different. In particular, the time intervals on a speeding rocket will appear to be longer to you than to the people on the ship.

Understanding why and how time varies

How does time dilation happen? To get the story, say that time is measured on a speeding rocket with a "light clock," as Figure 12-2 shows, so that every tick of the clock has a light ray traveling from one mirror to another and then back again.

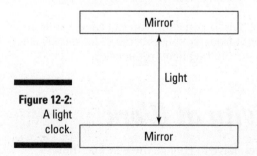

Figure 12-2:
A light
clock.

Now take a look at the situation from the point of view of an observer on the rocket, at the top of Figure 12-3, and from your point of view on Earth, at the bottom of Figure 12-3. To the observer in the rocket, light is just bouncing between the mirrors, a distance D, and each tick of the clock takes $2D/c$ seconds (the time for light to make it from one mirror to the other and back again). So for the observer on the rocket, call the time interval between ticks Δt_o.

The time interval measured from a reference frame at rest with respect to the event, Δt_o, has a special name: the *proper time interval*. So when the clock is on the rocket, the time between ticks is a proper time interval (the event is in the same reference frame as the measurement is made in).

But things are different from your point of view on Earth. Although the light ray is traveling the distance D between the mirrors, the rocket is moving forward a distance L, as you can see in the bottom of Figure 12-3. So the light ray has to travel a longer distance, s (where $s = [D^2 + L^2]^{1/2}$) — not just D), to strike the other mirror. And the light takes longer to make that longer trip, so the time you measure, Δt, is longer than the time measured on the rocket, Δt_o. In other words, distance equals speed times time, so if light speed remains constant, then time has to increase to give you a greater distance.

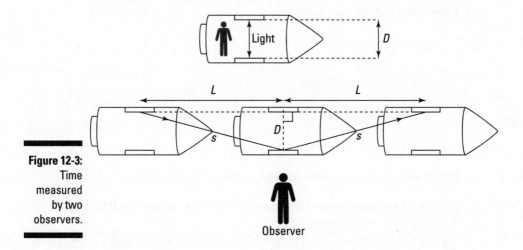

Figure 12-3:
Time
measured
by two
observers.

Look at this with a little math to relate Δt_o (the time on the rocket) and Δt (the time you measure). Start with the distance you observe the light to travel from one mirror to the other and back again, $2s$. Note that

$$2s = 2(D^2 + L^2)^{1/2}$$

Now you need to get some time into this equation. Note that the distance L is just the distance the rocket goes in the time you measure, Δt multiplied by the speed of the rocket from your point of view, which you call v, divided by 2. So $L = v\Delta t/2$, which means you can write the following:

$$2s = 2\left(D^2 + L^2\right)^{\frac{1}{2}}$$

$$= 2\left(D^2 + \left(\frac{v\Delta t}{2}\right)^2\right)^{\frac{1}{2}}$$

Now you need to get clever. Note that the distance $2s$ is the distance light travels in the time interval Δt — that is, $c\Delta t$. So this becomes

$$c\Delta t = 2\left(D^2 + \left(\frac{v\Delta t}{2}\right)^2\right)^{\frac{1}{2}}$$

Squaring and solving for Δt gives you the change in time:

$$\Delta t = \frac{2D}{c}\frac{1}{\left(1-\frac{v^2}{c^2}\right)^{\frac{1}{2}}}$$

But hang on a minute — $2D/c$ is the time the observer on the rocket measures for each tick of the light clock:

$$\Delta t_{\text{o}} = \frac{2D}{c}$$

So replace $2D/c$ to finally get

$$\Delta t = \frac{\Delta t_{\text{o}}}{\left(1-\frac{v^2}{c^2}\right)^{\frac{1}{2}}}$$

Here's a list of the variables to help you keep things straight:

✔ *Δt:* The time measured by an observer who is in motion with respect to the event being measured

✔ *Δt_{o}:* The time measured by an observer at rest with respect to the event being measured

✔ *v:* The relative speed of the two observers

✔ *c:* The speed of light in a vacuum

Time dilation is not just for light clocks — it means that time itself slows down. So the time measured by *any* clock in the rest frame is dilated when viewed by an observer who is moving.

Notice that when v is very much less than c ($v \ll c$), the time-dilation equation becomes

$$\Delta t = \Delta t_{\text{o}} \quad (v \ll c)$$

In other words, at low speeds, time dilation is not noticeable (which is why most physicists before Einstein had nothing to say about it). But as v approaches c, Δt becomes much larger than Δt_0.

Doing time-dilation calculations

Say that the rocket is moving at 0.95 times the speed of light, or $0.95c$. Then suppose a clock on the rocket measures 1.0 seconds between successive ticks. How long is the time between ticks, measured in your reference frame? The time-dilation equation comes to the rescue here:

$$\Delta t = \frac{\Delta t_0}{\left(1 - \frac{v^2}{c^2}\right)^{1/2}}$$

In this case, Δt_0 is 1.0 seconds (recall that Δt_0 is the time measured in the same reference frame as the event being measured), so

$$\Delta t = \frac{1.0 \text{ s}}{\left(1 - \frac{v^2}{c^2}\right)^{1/2}}$$

And $v = 0.95c$. After you square the velocity, the c^2 terms cancel out in the denominator, so you have the following:

$$\Delta t = \frac{1.0 \text{ s}}{\left(1 - 0.95^2\right)^{1/2}} \approx 3.2 \text{ s}$$

So on the ground, you measure the rocket's clock as taking 3.2 seconds between ticks, not 1 second. Pretty cool.

Note that the time dilation is noticeable because the relative speeds of the observers is so great: $0.95c$. What if instead of a rocket, the clock had been on a jet? Suppose the relative speed is only about 550 miles per hour, or about $(8.2 \times 10^{-7})c$. In that case, you'd have this time dilation:

$$\Delta t = \frac{1.0 \text{ s}}{\left(1 - \left(8.2 \times 10^{-7}\right)^2\right)^{1/2}} \approx 1.0000000000003 \text{ s}$$

In other words, you'd have to wait about 100,000 years before the time dilation between you and a clock on the jet amounted to 1 second. That's a long time to keep your stopwatch going.

Red light, blue light: Shifting light frequencies

Because of the Doppler effect (see Chapter 7), the pitch of sound waves depends on the motion of the source and the listener through the air. For instance, the siren on a police car sounds higher pitched as the car races toward you and lower pitched as it speeds away.

Although light is traveling in a vacuum, you still get a change in frequency. Imagine that a spaceship is traveling very quickly with its headlights on. As the ship travels toward you, you see a higher frequency and shorter wavelength (this is sometimes called a *blueshift*). If the ship travels away from you, then you see a lower frequency and longer wavelength (this is sometimes called a *redshift*).

The idea of time dilation can help you understand how this works. For the people on the spaceship, traveling at speed u, the light has a frequency f_0 and therefore a period $T_0 = 1/f_0$. What frequency of light do you see? Call the frequency and period of the light you see f and T (which equals $1/f$). As you see it, a peak of the light wave leaves the ship and travels a distance of cT in one cycle. In this time, you see the spaceship move a distance uT before emitting the next peak. So you see a wavelength of $(c - u)T$. Of course, like everyone else, you observe the light moving at speed c. Because

you know the relationship between wavelength speed and frequency ($c = \lambda f$), you can write the frequency that you observe as

$$f = \frac{c}{(c-u)T}$$

This is just the result you'd have with the Doppler effect for sound, but this time you have a difference: time dilation. You know that the period of the light you observe, T, is the time-dilated version of the proper time period T_0, which equals $1/f_0$. So you can write

$$f = \frac{c\left(1-\left(\frac{u}{c}\right)^2\right)^{\frac{1}{2}}}{(c-u)T_0} = \frac{c\left(1-\left(\frac{u}{c}\right)^2\right)^{\frac{1}{2}}}{(c-u)}f_0$$

Then if you do a little algebra to simplify, this becomes

$$f = \left(\frac{c+u}{c-u}\right)^{\frac{1}{2}}f_0$$

So you see that as the ship travels toward you, the top of the fraction is larger than the bottom, which gives you a larger frequency. The light may appear blue because it's shifted toward the violet/blue end of the visible light spectrum. If the ship travels away, you see a smaller frequency, making the light look red.

This all has some consequences for space travel. Given the great distances between stars, you may think you have no hope of reaching the stars, even if your rocket were going $0.99c$. But thanks to time dilation, time on board the rocket would pass much more slowly than an observer on Earth would measure.

Say, for example, that you have your heart set on visiting a star 10 light-years from Earth (a *light-year* is the distance light travels in one year, so it's c times the number of years). At $0.99c$, an external observer would say the trip would take

$$\Delta t = \frac{10c \text{ years}}{0.99c} \approx 10.1 \text{ years}$$

So you may think it'd take you 10.1 years to reach the star. But on the rocket, where the event is your aging from second to second, time passes much more slowly. In particular, if

$$\Delta t = \frac{\Delta t_o}{\left(1 - \dfrac{v^2}{c^2}\right)^{1/2}}$$

then

$$\Delta t_o = \Delta t \left(1 - \frac{v^2}{c^2}\right)^{1/2}$$

So the time that passes aboard the rocket would be

$$\Delta t_o = 10.1(1 - 0.990^2)^{1/2} \approx 1.4 \text{ years}$$

So although it looks to an observer on Earth that the trip would take 10.1 years, to you on the rocket, only 1.4 years would pass. Isn't physics wonderful? So you don't have to be too disappointed that you can't travel faster than light. If you get very close to c, then you can travel many light years in just a short time.

Packing it in: Length contraction

As if time dilation wasn't enough, special relativity also tells you that lengths get contracted at high speeds, an outcome of the finite speed of light. So although you may think that reaching stars is now possible due to time dilation, you'll have to put up with being only 2 inches wide. Just kidding — to you on the rocket, lengths would appear to be normal. But observers measuring the same events that you do (events happening in your rocket) would see length being contracted.

Looking at why and how lengths contract

Suppose you want to examine the length of a spaceship moving at the speed of light. You expect that the people on the ship and people observing from Earth will disagree about the spaceship's length, but you know that they *must* agree on the speed of light.

Length equals speed times time, so you can figure out the spaceship's length by first measuring the time that it takes a ray of light to cover a distance; then using the time-dilation equation (see the preceding section), you can find the lengths that each observer sees.

The time-dilation equation involves the time between events that happen at the same place in the *rest frame* (Δt_0), so here's how you set up the measurement: You plan to send a ray of light from the rear of the spaceship, reflect it from the front, and let it return to the same place. For the people on the ship, this takes a time

$$\Delta t_0 = \frac{2L_0}{c}$$

because the light ray travels a distance equal to twice the length of the ship as measured on it, L_0.

What time does an observer on Earth measure for this path of the light ray? You already expect that he'll observe the ship to be of a length L, which may be different from L_0. But he also sees the ship moving with speed v, so as the light ray travels from the rear of the ship to the front, the ship will have moved; therefore, the ray has to travel a distance slightly longer than L. On the return journey, the light ray has to travel a distance slightly shorter than L because of the movement of the ship. When you work out how much time this whole path would take a light ray at speed c, you get the following:

$$\Delta t = \frac{L}{c-v} + \frac{L}{c+v} = \frac{2L}{c\left(1-\dfrac{v^2}{c^2}\right)}$$

Now you can use the time-dilation equation to relate L and L_0. Here's how times are related by the time-dilation equation:

$$\Delta t = \frac{\Delta t_0}{\left(1-\dfrac{v^2}{c^2}\right)^{1/2}}$$

So if you plug the values of Δt and Δt_0 for the light-ray path into this equation, you get

$$\frac{2L}{c\left(1-\dfrac{v^2}{c^2}\right)} = \frac{\dfrac{2L_0}{c}}{\left(1-\dfrac{v^2}{c^2}\right)^{1/2}}$$

Then rearrange the equation to get the equation for length contraction:

$$L = L_0\left(1-\frac{v^2}{c^2}\right)^{1/2}$$

Cool — that relates the lengths measured by the two observers. Here's what all the variables stand for:

- ✔ **L:** The length measured by an observer who is in motion with respect to the distance being measured
- ✔ **L_o:** The length measured by an observer at rest with respect to the distance being measured
- ✔ **v:** The relative speed of the two observers
- ✔ **c:** The speed of light in a vacuum

Notice that the factor $(1 - v^2/c^2)^{1/2}$ is always less than 1 (because objects can never actually reach the speed of light, although they can come very close). That means that the Earth observer sees a length measured on the rocket contract. So even if the rocket is 100 meters long, the Earth observer may see it as only 10 meters long, depending on the relative speed of the two observers.

Length contraction happens only along the direction of travel. Distances perpendicular to the direction of travel are not affected. In other words, if the rocket is 100 meters long and 20 meters wide as measured by an observer on the rocket and it's going so fast past the Earth that an observer on the Earth measures it as being only 10 meters long, the Earth observer would still see the rocket as being 20 meters wide. (The Earth observer would, no doubt, think it's a pretty funny-looking rocket.)

Trying some length-contraction calculations

Say that the relative speed between the rocket ship and Earth is 0.99c. The observers on the rocket and the Earth have noticed that they aren't agreeing on lengths of items on the rocket ship. So the rocket observer holds up a meter stick (its length being in the direction of the rocket's travel away from Earth) and asks the Earth observer to measure it, expecting the Earth observer to report exactly 1.00 m.

You know that length along the direction of relative travel will contract as measured by the Earth observer, and you whip out your handy formula:

$$L = L_o \left(1 - \frac{v^2}{c^2} \right)^{\frac{1}{2}}$$

Here, L_o is the *proper length,* the length measured by the rocket observer, and L is the length measured by the Earth observer, who is in motion with respect to the proper length.

Plugging in the numbers gives you

$$L = (1.00)\left(1 - \frac{(0.99c)^2}{c^2}\right)^{\frac{1}{2}} \approx 0.14 \text{ m}$$

So you may hear the Earth observer radio the rocket observer and say, "Hey buddy — somebody sold you a defective meter stick! That one's only 14 centimeters long. Here, take a look at mine."

And the Earth observer holds up a meter stick, also in the direction of travel of the rocket. A laugh comes back over the radio, "Hey pal, you're the one with the defective meter stick! That one's only 14 centimeters long. Don't they teach you anything on Earth?"

In other words, when the Earth observer held up a meter stick, the meter stick measured 1.00 meters — on Earth. To the rocket observer, the meter stick is in motion (so the Earth measurement is the *proper length, L_o*, because it's made at rest with respect to the meter stick), and the rocket observer sees it with a contracted length.

L_o is always the proper length — the length measured at rest with respect to the thing you're measuring — and t_o is the proper time — the time measured at rest with respect to whatever you're timing.

Pow! Gaining momentum near the speed of light

As if time dilation and length contraction weren't enough, special relativity also affects momentum. From Physics I, you find out that *momentum* is mass multiplied by velocity, and its symbol is *p*:

$p = mv$

When you roll a pool ball around or throw a baseball, it has momentum — it's the *oof!* factor that makes moving things hard to stop. (Note that momentum is a vector, of course, but I talk only in terms of its magnitude here.)

Special relativity has something to say about momentum. In particular, special relativity gets its $(1 - v^2/c^2)^{1/2}$ factor into the momentum mix like this:

$$p = \frac{mv}{\left(1 - \frac{v^2}{c^2}\right)^{\frac{1}{2}}}$$

Here's how the variables relate:

- *m:* The mass of the moving object
- *v:* The speed of the object that you measure
- *c:* The speed of light in a vacuum
- *p:* The object's momentum

Note that because $(1 - v^2/c^2)^{1/2}$ is always less than 1, the relativistic momentum is always greater than the classical momentum *(mv)*, but the difference isn't noticeable at slower speeds. So you can safely assume that the momentum and the relativistic momentum of a pool ball speeding across the table to a pocket are about the same.

The difference starts becoming noticeable at higher speeds, of course. About the highest speeds that humans have been able to give to objects with mass are those reached in *particle accelerators,* which are those rings or linear tracks that physicists use to get particles like electrons moving at relativistic speeds.

The speeds of electrons in those accelerators is very fast, pretty close to the speed of light. How close? At the Stanford Linear Accelerator Center (SLAC) in California, electrons are routinely goosed to speeds of 0.9999999997*c*. That fast enough for you? Physicists can get pretty speedy when they want to.

Classically, such electrons should only have a momentum of

$$p = mv$$
$$= (9.11 \times 10^{-31} \text{ kg})(0.9999999997c)$$
$$= 2.7 \times 10^{-22} \text{ kg-m/s}$$

But Einstein tells you that the electrons' momentum is really

$$p = \frac{mv}{\left(1 - \frac{v^2}{c^2}\right)^{1/2}} \approx 1.0 \times 10^{-17} \text{ kg-m/s}$$

Note that you may not be able to put all the digits of the electrons' speed on your calculator, but if you look at Figure 12-4, you can see that the extra factor that appears in relativistic momentum becomes larger and larger as the speed approaches the speed of light.

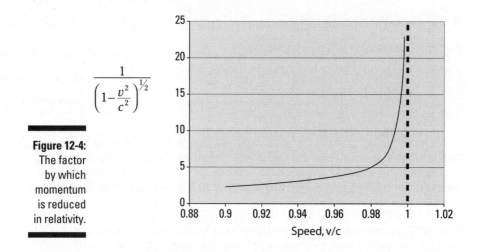

$$\frac{1}{\left(1-\dfrac{v^2}{c^2}\right)^{\frac{1}{2}}}$$

Figure 12-4:
The factor
by which
momentum
is reduced
in relativity.

So although still tiny, the electrons' momentum is a factor of

$$\frac{1.0\times10^{-17}\ \text{kg m/s}}{2.7\times10^{-22}\ \text{kg m/s}} \approx 3.7\times10^{4}$$

That is, at relativistic speeds, the electrons' momentum is 37,000 times what the momentum would be if the classical momentum held.

Note that different inertial reference frames can move at various speeds with respect to each other — and that means that momentum is not conserved between inertial reference frames. For example, a pool ball traveling slowly on the pool table may be seen as traveling very fast by a rocket-based observer — which means that the momentum you and the rocket observer measure would be different.

Here It Is! Equating Mass and Energy with E = mc²

Perhaps the most startling result of special relativity, and the basis of the world's most well-known physics equation, is that mass and energy are equivalent. Strictly speaking, that means that when you add energy to an object, it's the same as adding mass.

So what's the world's most famous physics equation? Here it is:

$$E = \frac{mc^2}{\left(1 - \frac{v^2}{c^2}\right)^{1/2}}$$

Hmm, perhaps not exactly what you were expecting. That's the complete version of the equation, which includes energy due to the relative motion between you and the mass (kinetic energy). You were perhaps expecting this:

$$E_o = mc^2$$

That's the same equation, with $v = 0$ (so you're at rest with respect to the mass involved, which is why it's E_o, not just E). E_o is called a mass's *rest energy*.

In this section, you work with both versions of the formula, noting both rest energy and kinetic energy. You also see how to include potential energy in the equation.

An object's rest energy: The energy you could get from the mass

An object's *rest energy*, E_o, is energy-equivalent of a mass at rest if it were converted to pure energy. Einstein's famous equation tells you that an amount of mass m has an equivalent amount of energy E_0, given by $E_0 = mc^2$.

Physicists can observe the equivalence of mass and energy in experiments where particles called *neutral pions* disappear. So what happens to the conservation of mass? Well, when the pion disappears, it leaves light, which has an energy that's equivalent to the missing mass (times c^2). Energy and mass are two sides of the same coin, and together they're conserved.

In this section, you see how mass can turn into pure energy.

Converting between mass and energy

How could you tell that $E_o = mc^2$ experimentally? Well, you could do it if you had two pool balls, one made of matter and the other made of *antimatter* (the opposite of matter, where electrons have positive charge and protons have negative charge). If you brought the balls together, they'd flash into pure energy (released as light), and the energy released would be the same as mass of the two pool balls multiplied by c^2.

In fact, physicists working at particle accelerators convert mass into energy and back again every day. You may be interested to know (sci-fi fans take note) that physicists are creating antimatter here on Earth every day of the week — in tiny amounts. Particle accelerators regularly create *positrons* — the antimatter equivalent of electrons. Positrons have a positive charge but the same mass as a normal electron. So here's how the conversion between mass and energy works:

- **Changing mass to energy:** When you bring an electron and a positron together, there's a very, very tiny explosion, and two photons of high energy (gamma rays) are created. That's the conversion of mass into pure energy. When you measure the energy of the created photons, sure enough, the theory of special relativity is right.

- **Changing energy to mass:** Conversely, two gamma rays, colliding head-on, can do the reverse and end up as an electron and a positron. That's the conversion of pure energy into mass.

The masses of electrons and positrons are miniscule, but if you convert larger masses into energy, the amount of energy created can be enormous, because all the mass is converted into energy. By contrast, in a nuclear explosion, only about 0.7 percent of the mass involved is converted into energy.

Powering up: Finding the energy in a jar of baby food

Now check out some numbers. Suppose you have a jar of baby food with a mass of 46 grams. If you were to convert all the food in the jar into pure energy (don't try this at home!), how long would it keep a 100-watt bulb going?

First, find the energy that'd be released by converting the jar of baby food into energy, using the rest-energy equation:

$$E_0 = mc^2$$

Plugging in the numbers gives you

$$E_0 = (0.046 \text{ kg})(3.0 \times 10^8 \text{ m/s})^2 = 4.1 \times 10^{15} \text{ J}$$

That's a lot of joules. How long would it keep a 100-watt bulb burning? Well, the time the bulb would keep going is

$$\text{Time} = \frac{\text{energy}}{\text{power}}$$

So that's

$$\text{Time} = \frac{4.1 \times 10^{15} \text{ J}}{100 \text{ W}}$$
$$= 4.1 \times 10^{13} \text{ s}$$

And that's a mere 1.3 million years. Not bad for a little jar of baby food.

Shrinking the sun: Turning mass into light

Of course, you may not convert jars of baby food into pure energy every day. Here's another example — the sun is getting lighter. The sun is losing mass every second by converting its mass into energy, which it beams away as sunlight. You don't simply have little hunks of the sun flying into space; the light particles, *photons,* are massless — they've become pure energy.

How much mass is the sun losing every second? Well, if the sun were a light bulb, it would be a 3.92×10^{26}-watt light bulb — that is, a 392,000,000,000,000,000,000,000,000-watt light bulb. So in 1 second, the sun loses this much energy (a watt is 1 joule per second):

$$\Delta E_o = 3.92 \times 10^{26} \text{ J}$$

And because $E_o = mc^2$, the amount of mass the sun loses is

$$\Delta m = \frac{\Delta E_o}{c^2} = \frac{3.92 \times 10^{26} \text{ J}}{\left(3.0 \times 10^8 \text{ m/s}\right)^2} \approx 4.36 \times 10^9 \text{ kg}$$

That's about the mass of about 47 aircraft carriers, which is a lot of mass to burn every second.

An object's kinetic energy: The energy of motion

From the complete version of Einstein's equation relating mass and energy, you know that the total energy of an object in motion is the following:

$$E = \frac{mc^2}{\left(1 - \dfrac{v^2}{c^2}\right)^{1/2}}$$

What's this *total energy* made up of? You know that for an object at rest, the energy is $E_o = mc^2$. The remainder of the total energy is *kinetic energy* — the energy of motion:

Total energy = rest energy + kinetic energy

So this means that the kinetic energy of an object is

Kinetic energy = total energy – rest energy

Therefore, the kinetic energy, *KE*, is equal to

$$KE = \frac{mc^2}{\left(1 - \frac{v^2}{c}\right)^{1/2}} - mc^2$$

$$= mc^2 \left(\frac{1}{\left(1 - \frac{v^2}{c^2}\right)^{1/2}} - 1 \right)$$

Relating the relativistic formula to one from classical mechanics

The relativistic formula for kinetic energy doesn't look very much like the old familiar nonrelativistic version for the kinetic energy here:

$$KE = \frac{1}{2}mv^2$$

But actually, the relativistic version reduces to the nonrelativistic form when *v* is much less than *c* (that is, $v \ll c$). That's because at low speeds, *v* is small, and you can expand the factor $1/(1 - v^2/c^2)^{1/2}$ to get the following (this is a Taylor expansion):

$$\frac{1}{\left(1 - \frac{v^2}{c^2}\right)^{1/2}} = 1 + \frac{1}{2}\left(\frac{v^2}{c^2}\right) + \frac{3}{8}\left(\frac{v^4}{c^4}\right) + \dots$$

Because $v \ll c$, you can ignore the third and higher terms here, so to a good approximation, you can say the following:

$$\frac{1}{\left(1 - \frac{v^2}{c^2}\right)^{1/2}} \approx 1 + \frac{1}{2}\left(\frac{v^2}{c^2}\right) \qquad (v \ll c)$$

Plug this result into the equation for kinetic energy and simplify to get the familiar version of the *KE* equation from mechanics:

$$KE = mc^2 \left(\frac{1}{\left(1 - \frac{v^2}{c^2}\right)^{1/2}} - 1 \right) \qquad (v \ll c)$$

$$KE \approx mc^2 \left(1 + \frac{1}{2}\left(\frac{v^2}{c^2}\right) - 1 \right)$$

$$KE \approx mc^2 \left(\frac{1}{2}\right)\left(\frac{v^2}{c^2}\right)$$

$$KE \approx \frac{1}{2} mv^2$$

So all the time that you've been using this equation for kinetic energy, you've really been using an approximation to the relativistic equation:

$$KE = mc^2 \left(\frac{1}{\left(1 - \frac{v^2}{c^2}\right)^{1/2}} - 1 \right)$$

Who knew?

One consequence of the kinetic-energy equation is the conclusion that objects with mass cannot reach the speed of light (sorry). That's because as you get closer and closer to the speed of light, the denominator here approaches zero, which makes the kinetic energy approach infinity. And it'd take an infinite amount of work to give something an infinite kinetic energy. However, an infinite amount of work isn't available (and imagine your energy bill if it were!), so the conclusion is that giving the speed of light to an object that has mass is impossible. The only way out is to set *m* to zero, which makes the kinetic-energy equation meaningless.

Plugging in some numbers to find the KE

Say that you see a rocket of mass 10,000 kilograms passing overhead at 0.99*c*. What are its total energy and kinetic energy? Its total energy is

$$E = \frac{mc^2}{\left(1 - \frac{v^2}{c^2}\right)^{1/2}}$$

$$= \frac{(10,000 \text{ kg})(3.0 \times 10^8 \text{ m/s})^2}{\left(1 - 0.99^2\right)^{1/2}}$$

$$\approx 6.4 \approx 10^{21} \text{ J}$$

By contrast, the kinetic energy equals

$$KE = mc^2\left(\frac{1}{\left(1-\frac{v^2}{c^2}\right)^{\!1\!/\!2}}-1\right)$$

$$= 10,000 \text{ kg}\left(3.0\times10^8 \text{ m/s}\right)^2\left(\frac{1}{\left(1-0.99^2\right)^{\!1\!/\!2}}-1\right)$$

$$\approx 5.5\times10^{21} \text{ J}$$

Skipping PE

Einstein's total-energy equation is just the sum of an object's rest energy and its kinetic energy — it ignores potential energy. If you want to include potential energy in this equation, you have to add it. For example, if an object has potential energy by being at a certain height in the Earth's gravitational field, you can add that potential energy, *mgh,* to the total-energy equation:

$$E = \frac{mc^2}{\left(1-\frac{v^2}{c^2}\right)^{\!1\!/\!2}} + mgh$$

where *m* is the object's mass, *g* is the acceleration due to gravity on the Earth's surface, and *h* is the height of the object.

The rest energy of a particle is not a kind of potential energy; it's the energy a massive particle has at rest, just because it has mass. Potential energy comes from a particles' position in a field of any kind — gravitational, electric, and so on.

New Math: Adding Velocities Near Light Speed

The speed of light postulate (from the earlier section "Looking at special relativity's postulates") says that the speed of light in a vacuum, *c,* always has the same value in any inertial reference frame, no matter how fast the observer and the light source are moving with respect to each other.

Suppose you have two rockets traveling toward each other, each traveling at 0.75c as measured by an observer on Earth. Would an observer in one rocket see the other rocket approach at a speed of 0.75c + 0.75c? No, because that's 1.5c, which isn't possible.

As you can see, special relativity has to make some provisions for adding velocities so you don't end up with velocities greater than c. This formula allows you to find the sum of the speeds, bringing you closer to the limit c without passing it.

Take a look at the situation in Figure 12-5. There, a rocket is traveling past an observer on Earth at speed v_{rocket} (the observer on Earth sees the same speed for the rocket, only in the opposite direction). Now say that the crew in the rocket recently did some work on the outside of the rocket, and they sloppily left a wrench outside. That wrench is traveling away from the rocket — observers on the rocket measure the speed of the wrench as v_o.

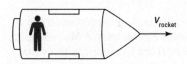

V_{rocket}

Figure 12-5:
A rocket
and wrench
moving at
speeds
nearing c.

v_0 measured on rocket

Observer

What does the observer on Earth at the bottom of the figure measure the speed of the wrench as? That observer measures the speed of the wrench as v, so is the following equation true?

$$v = v_o + v_{rocket} \ ?$$

Nope, because you could conceivably get a speed greater than the speed of light by just adding two speeds together this way. Instead, Einstein says that the speed of the wrench as measured on Earth is

$$v = \frac{v_o + v_{rocket}}{1 + \dfrac{v_o v_{rocket}}{c^2}}$$

where

- v is the speed of the wrench as measured on Earth
- v_o is the speed of the wrench as measured on the rocket
- v_{rocket} is the speed of the rocket relative to the Earth

Say that the speed of the rocket relative to the Earth is $0.75c$, and the speed of the wrench with respect to the rocket, as measured on the rocket, is also $0.75c$. Instead of simply adding these two speeds to get $1.5c$, you use the relativistic equation. Putting in the numbers, $v_o = v_{rocket} = 0.75c$, gives you

$$v = \frac{0.75c + 0.75c}{1 + 0.75^2}$$
$$= \frac{0.75c + 0.75c}{1 + 0.75^2}$$
$$\approx 0.96c$$

The law for adding velocities in relativity was observed even before Einstein discovered relativity. In 1851, Hippolyte Fizeau used a Michelson interferometer with flowing water between the mirrors to compare the velocity of light in stationary water to the velocity of light in moving water (an *interferometer* separates light beams so researchers can observe interference patterns and make conclusions about the light waves — see Chapter 11). The expectation was that the velocities of light and the water should simply add up. However, the velocities added relativistically. Nobody could explain the result of this experiment until Einstein came along and showed how people needed to rethink their most basic ideas of space and time.

Chapter 13

Understanding Energy and Matter as Both Particles and Waves

*W*hat is *matter?* That's a question that physicists have long been asking. And they've come up with some surprising answers. Everybody knows what electrons and protons are, right? They're tiny particles that orbit around each other to form atoms, and they're the fundamental building blocks of matter. But it turns out that the particle nature of electrons and protons and all matter isn't quite right: Such particles can also act like waves. That sort of challenges the imagination — how can a baseball act as anything other than an object? How can matter act as a *wave,* a traveling disturbance that transfers energy? That's the kind of question you look at in this chapter.

On the other hand, physicists have also been asking questions about light, which is known for its wave qualities. Chapter 11 covers how light works as a wave — for instance, how light passing through a pair of slits can interfere with itself and cause constructive and destructive interference. But light can show particle-like qualities, too — you've heard of *photons,* which are particles of light. So what is light? Waves or particles? The answer to that question is *both:* Light exhibits both particle and wave qualities, depending on what you're measuring.

In this chapter, you look at the particle nature of light, the wave nature of electrons, and the experiments that suggested the relationships between energy and matter. All this ties in nicely with Einstein's idea that mass and energy are equivalent, $E = mc^2$, which I cover in Chapter 12. The result is a more complete picture of waves, particles, energy, and momentum. I start this chapter — as physicists started historically — by talking about the particle nature of light.

Blackbody Radiation: Discovering the Particle Nature of Light

The first experiment that showed how light could act like particles had to do with explaining the radiation spectrum of light that every object emits.

Blackbody radiation is the radiation from an ideal surface, which absorbs any wavelength of radiation incident upon it. Physicists studied blackbody radiation extensively by experiment and knew much about it by 1900, but little was understood until some quite revolutionary changes in physics. Not only did the problem of blackbody radiation suggest the particle nature of light, but it also led to the field of quantum physics. In this section, I explain the experimental results that hinted that light is more than just a wave.

Understanding the trouble with blackbody radiation

A glowing piece of charcoal, temperature about 1,000 kelvin, emits a cherry-colored light that you can see. And although people, who have a temperature of about 310 kelvin, don't glow in the visible spectrum, they emit infrared light, which is visible to night scopes.

Physicists studied the spectrum of that light and found that it varied by the temperature of the object in question. Figure 13-1 shows the spectrum of emitted light — intensity versus wavelength — from a perfect blackbody (*intensity* is the amount of energy radiated by the wave per unit area per unit time, as I explain in Chapter 8). A *perfect blackbody* is simply an object, any object, that emits as much light as falls upon it from its environment.

The thing that puzzled physicists in the early part of the 20th century was the shape of the spectrum. As the temperature of the blackbody increased, the wavelength of the light emitted with the highest intensity decreased, creating a spectrum with the characteristic shape that you see in the figure. Physicists advanced plenty of theories as to just how a blackbody worked, but each theory was incomplete — at best, it could match only one part of the spectrum, at low wavelengths or at high wavelengths. But no one could give a satisfactory theoretical model of how blackbodies produced exactly the spectrum you see in Figure 13-1.

One attempt at an explanation of the blackbody spectrum only in terms of waves was made by Lord Rayleigh (John William Strutt, third Baron Rayleigh). His theory produced a prediction for the blackbody spectrum that fit quite well at large wavelengths; however, it had quite a serious problem — it predicted that a blackbody would radiate with infinite power! The predicted spectrum increased to infinity at shorter and shorter wavelengths. All attempts to explain the blackbody spectrum only with waves had similar problems.

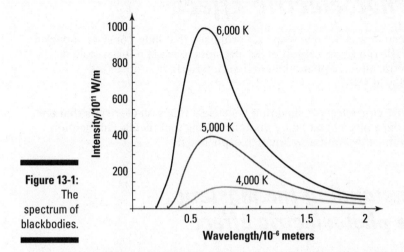

Figure 13-1:
The
spectrum of
blackbodies.

Being discrete with Planck's constant

Max Planck, a German physicist, came along with a truly radical idea: He said that you have to consider the blackbody as a collection of many atomic oscillators, each of which emits radiation. The radical part of Planck's idea was that such atomic-sized oscillators could only emit energies of

$$E = nhf \qquad n = 0, 1, 2, 3, \ldots.$$

where n is a positive integer, f is the frequency of the oscillator, and h is a constant known as Planck's constant:

$$h = 6.626 \times 10^{-34} \text{ J} \cdot \text{s}$$

That is, each atomic oscillator could only radiate energies that were discrete, that were multiples of hf. Other energies were not allowed. Today, when only certain energy states are allowed, you say that the system is *quantized*. That was the beginning of quantum physics. (You can find out more about quantum physics in my book *Quantum Physics For Dummies* [Wiley].)

The fact that energy could be emitted only with certain energies meant that not only were the atomic oscillators quantized but the emitted light was, too. In other words, light generated by a blackbody exists in discrete *quanta*, with only certain energies allowed. That contradicted the classical picture of light as a continuous spectrum of all possible wavelengths. Planck's result implied the particle-like nature of light, with each particle of light having its own allowed energy.

Light Energy Packets: Advancing with the Photoelectric Effect

Albert Einstein was the one who first proposed that light consists of packets of energy. He did so as a result of his attempts to explain the so-called photoelectric effect, a phenomenon Heinrich Hertz first observed accidentally in 1887.

The *photoelectric effect* is called that because it relies on electrons that are ejected from a piece of metal by photons hitting that metal. This section describes the effect and how Einstein explained it.

Understanding the mystery of the photoelectric effect

An experimental apparatus to measure the photoelectric effect appears in Figure 13-2. Here's how it works: Electrons are normally trapped in the metal, attracted by the positive charge of the metal atoms' nuclei. Even when a voltage is applied across the gap in the figure (between the metal sheet and the collector), the electrons are bound so tightly to the metal that they don't leave the surface.

But when light shines on the sheet of metal, the light interacts with the atoms of metal, exciting them. Under certain circumstances, this light can cause electrons to break free from the surface of the metal. When the light gives the electrons the energy they need to leave the surface of the metal, they're kicked out.

Those emitted electrons then travel to a positive plate, called the *collector*, as Figure 13-2 shows. The metal plate and collector are in a vacuum (inside a glass bell-jar or tube) to minimize the collisions of the electrons with the atoms of the air, which would complicate matters. Because the electrons travel from one metal plate to another, current flows (although a very small current), which can be measured by the meter at the bottom of the figure. So when you shine light onto the metal, current flows. It's as simple as that.

To isolate the effect of the frequency of the incident light, researchers decided to shine monochromatic light (light of a particular frequency) on the metal plate. They could then study the effects of varying the frequency and intensity of this light separately.

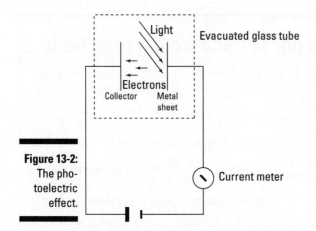

Figure 13-2:
The pho-
toelectric
effect.

Light would fall on the atoms of the metal, and physicists expected that when they used enough light waves, electrons would gather enough energy to be emitted. So classically, the more intensely the light shone on the metal, the more energy the emitted electrons should have. The assumption was that at very low levels of light, electrons would need some time to gather enough energy to be emitted. But that's not what happened. Here are two surprising findings:

✔ The energy of the emitted electrons turned out to be independent of the intensity of the light: If researchers doubled the amount of light, the electrons they saw didn't end up with any different energy when they were emitted.

✔ When researchers shone even low-intensity light on the metal, electrons started to be emitted immediately; it didn't take time for them to gather enough energy before being emitted.

Einstein to the rescue: Introducing photons

Einstein, building on the work published by Max Planck (see the earlier section "Being discrete with Planck's constant"), proposed that light was actually made up of discrete energy packets — today, you know those as *photons*. In particular, again following Planck, Einstein said that the energy of each photon is equal to

$$E = hf$$

where E is the energy of the photon, h is Planck's constant (6.626×10^{-34} joule-seconds), and f is the frequency of the photon.

Calculating photons per second from a light bulb

Consider an electric light bulb of 100 watts. How many photons does it emit per second? Can you even think in terms of such a question? Yes, you certainly can. But you need to know the energy that the light emits every second and the energy of every photon. Because a light bulb's spectrum more or less covers the visible spectrum, assume that the average visible wavelength of light from the bulb is green light (λ = 555 nanometers).

Okay, how much energy does the bulb emit per second? That's 100 watts, which is 100 joules per second, right? Not exactly. Incandescent lights are only about 2 percent efficient — that is, a 100-watt bulb emits only 2 joules of visible light per second.

Okay, how many photons are in 2 joules per second? To determine that, you have to find the energy of each photon. You're assuming the wavelength of the emitted light is green, on average λ = 555 nanometers. What's the

frequency? You can relate the speed of light *(c)* to its frequency *(f)* and wavelength (λ) like this:

$$f = \frac{c}{\lambda}$$

So the frequency of green light is

$$f = \frac{c}{\lambda} = \frac{3.0 \times 10^8 \text{ m/s}}{555 \times 10^{-9} \text{ m}} \approx 5.40 \times 10^{14} \text{ Hz}$$

That means the energy of one photon is

$$E_{photon} = (6.626 \times 10^{-34} \text{ J-s})(5.40 \times 10^{14} \text{ Hz})$$
$$\approx 3.58 \times 10^{-19} \text{ J}$$

So the number of emitted photons per second is

$$\frac{\text{Energy/second emitted}}{E_{photon}} = \frac{2 \text{ J/s}}{3.58 \times 10^{-19} \text{ J}}$$
$$\approx 6 \times 10^{18} \text{ photons/s}$$

So a 100-watt bulb emits about 6×10^{18} photons per second in the visible spectrum — a huge number.

Einstein's equation shows that the energy of each photon depends on light frequency. For the photoelectric effect, Einstein suggested that each electron absorbs one photon, so the energy of the emitted electrons depends on light frequency as well. More-intense light contains more photons, so intensity can affect the number of electrons emitted but not the energy.

Now that you're getting into photons, it's worth asking what their mass is. After all, the whole point here is that photons act like particles. So do they have mass? From Chapter 11, you know that the energy of something in motion is

$$E = \frac{mc^2}{\left(1 - \dfrac{v^2}{c^2}\right)^{1/2}}$$

So rearrange the equation to isolate the mc^2 term:

$$E\left(1-\frac{v^2}{c^2}\right)^{\frac{1}{2}} = mc^2$$

Now the $(1 - v^2/c^2)^{1/2}$ term is zero, because by definition for photons, $v = c$. The energy is not zero, but the product of E times zero is zero, so $mc^2 = 0$ — which means that m is zero. So the mass of photons is zero, nothing, nada.

Explaining why electrons' kinetic energy is independent of intensity

So how exactly did Einstein use photons to explain the photoelectric effect? He had two issues to explain here: the idea that the kinetic energy of the emitted electrons is independent of the light intensity and the fact that electrons are emitted immediately, even in low-intensity light. I discuss kinetic energy in this section and the immediate release of electrons in the next.

Classically, you'd expect electrons to be emitted by electromagnetic waves with a continuous spectrum — and the more intense the light, the faster the ejected electrons should be going. But that's not what happens. For a particular frequency of light, the ejected electrons have a particular kinetic energy — and even if you shine twice as much light on the metal, you don't get electrons with more kinetic energy (you do get more electrons, however).

The particle theory of photons explains the photoelectric effect by saying that the energy of each photon — and therefore the energy it can deliver to a single electron — is dependent only on its frequency. So instead of having electrons absorb energy by being bathed in continuous light, each electron absorbs one photon.

That's why the kinetic energy of the emitted electrons is independent of the light intensity: The intensity of the light determines only the number of photons, not their individual energy. It's the photon energy that determines the kinetic energy of the ejected electrons.

According to Einstein's photon model, each photon's energy goes into

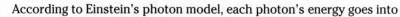

✔ The energy needed to pull an electron out of the metal

✔ The kinetic energy of that electron

The energy needed to pull electrons out of metal is called that metals' *work function,* or *WF,* so the energy of each photon, which is *hf,* is equal to the following:

$$hf = KE + WF$$

where *KE* is the kinetic energy of the ejected electron and *h* is Planck's constant (6.626×10^{-34} J-s).

This equation for the kinetic energy of the emitted electron tells the whole story: The kinetic energy of an emitted electron is just dependent on the frequency of the incoming photons, not their number, and the work function of the metal.

Explaining why electrons are emitted instantly

To describe the photoelectric effect, the second problem that Einstein had to solve was why electrons were emitted instantly when light — even low-intensity light — was shone on the metal.

Classically, you'd expect light intensity to have to build up enough energy to start ejecting electrons. But using Einstein's energy-packet theory, you don't need to wait until low-intensity light waves build up enough energy to emit electrons, because the light is actually made up of energy packets whose energy is dependent only on their frequency.

That means that as soon as you shine the light on the metal, you have photons that are energetic enough to eject electrons — no need to wait for the light to build up enough energy; each photon already has enough energy. Therefore, you still get electrons when you shine low-intensity light on metal; you just get fewer in number than when you shine more-intense light on the metal. Einstein triumphs again.

Einstein and the big prize

Everybody knows that Einstein won the Nobel Prize because of $E = mc^2$, right? Wrong. Einstein won the Nobel Prize in 1921 because of his work on the photoelectric effect. However, he was so prolific that he could've won it several times over.

Winning the big prize is quite a big deal. I have a friend at Cornell University who won the Nobel Prize in physics. They called him at 5 a.m. direct from Stockholm, and he was so excited that he didn't get any sleep for the rest of that night — or the next night, either.

Doing calculations with the photoelectric effect

Take a look at an example. Say that as a good physicist, you practice the photoelectric effect on the first metal you come across, which happens to be Mom's good set of silver spoons. When you shine your flashlight on the silver, do electrons pop out?

The work function of a metal is usually given in electron volts, eV, and 1 *electron volt* is the energy needed to move one electron through 1 volt of potential (you have to push the electron in order to do work on it, so you may think of this as pushing the electron toward the negatively charged plate of a parallel plate capacitor):

$1 \text{ eV} = 1.60 \times 10^{-19} \text{ J}$

(That's because work = $q\Delta V$, and q for an electron = 1.60×10^{-19} C, whereas $\Delta V = 1.0$ V).

The work function of silver *(WF)* is 4.72 eV, so you need that many electron volts to free an electron from the silver. So what's the frequency you need to start freeing electrons?

Converting 4.72 eV into joules gives you the energy needed to overcome the work function:

$E_{needed} = (4.72 \text{ eV})(1.60 \times 10^{-19} \text{ J/eV}) \approx 7.55 \times 10^{-19} \text{ J}$

Okay, so you need photons with an energy of 7.55×10^{-19} J. What frequency does that correspond to? You know that

$E_{photon} = hf$

where h is Planck's constant (6.626×10^{-34} joule-seconds) and f is the frequency of the photon. So you can rearrange the formula to say

$f = \dfrac{E_{photon}}{h}$

Because the energy of the photon needs to be at least 7.55×10^{-19} J, you find that the minimum frequency needed is

$f = \dfrac{7.55 \times 10^{-19} \text{ J}}{6.626 \times 10^{-34} \text{ J-s}}$

$\approx 1.14 \times 10^{15} \text{ Hz}$

Okay, so the light you shine on Mom's silver must have a frequency of at least 1.14×10^{15} hertz to eject electrons. What wavelength of light, λ, does that correspond to? Because $c = \lambda f$, you know that

$$\lambda = \frac{c}{f}$$

So the wavelength of light corresponding to the minimum frequency that you need is

$$\lambda = \frac{3.00 \times 10^8 \text{ m/s}}{1.14 \times 10^{15} \text{ Hz}} \approx 2.63 \times 10^{-7} \text{ m} = 263 \text{ nm}$$

So the light must have a wavelength equal to or shorter than 263 nanometers — and that's in the ultraviolet range, so your flashlight won't work for the trick.

Collisions: Proving the Particle Nature of Light with the Compton Effect

Even though Einstein had announced that light travels in energy packets, the particle-like nature of light wasn't fully accepted for several more years. What happened to change everybody's mind? In 1923, physicist Arthur Compton performed an experiment in which he bounced photons off electrons, showing that both electrons and photons were scattered by the collision. And if that doesn't prove the particle nature of photons, what would?

Compton sent beams of *X-rays* (that is, high-frequency photons) into targets made of graphite that had electrons at rest, just waiting to be hit. He observed that the photons were actually scattered by their collisions with electrons. He also noticed that the frequency of the scattered photons was lower than that of the incident photons, indicating that the photon had transferred some energy to the electron, which was initially at rest.

Not only do photons and electrons collide — they collide *elastically*, which means that both momentum and kinetic energy are conserved during the collision. In other words, the electron and photon bounce off each other in much the same way as billiard balls would. You can see a diagram of the scattering in Figure 13-3.

Energy is conserved when the electron is scattered by the photon. What's that look like? That means that

$$E_{\text{incident photon}} = E_{\text{scattered photon}} + KE_{\text{scattered electron}}$$

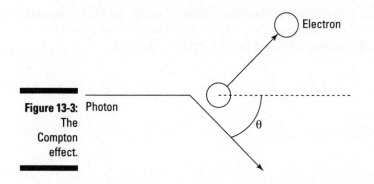

Electron

Photon

θ

Figure 13-3:
The
Compton
effect.

That is, the energy of the incident photon goes into the energy of the scattered photon and into the kinetic energy of the scattered electron (recall that the electron starts at rest). That's half of the picture — the energy-conservation half. How about the momentum-conservation half of the picture?

For a colliding photon and electron, the momentum-conservation equation looks similar to the energy-conservation equation, except that here you're dealing with momentum vectors, **p,** like this:

$$\boldsymbol{p}_{\text{incident photon}} = \boldsymbol{p}_{\text{scattered photon}} + \boldsymbol{p}_{\text{scattered electron}}$$

The momentum of a scattered electron is no problem — that's mv, or the following in relativistic form (see Chapter 12 for details on special relativity and what happens at speeds near the speed of light):

$$p = \frac{mv}{\left(1 - \frac{v^2}{c^2}\right)^{1/2}}$$

where p is the momentum of an object, m is the mass of the object, and v is the speed of the object.

So that's okay for an electron. But what about the momentum for a photon, which doesn't have any mass? Does that automatically mean that photons have no momentum? No, as the Compton effect demonstrates. The energy of a relativistic particle looks like this (also from Chapter 12):

$$E = \frac{mc^2}{\left(1 - \frac{v^2}{c^2}\right)^{1/2}}$$

This equation contains the mass of the particle, too. So are you stuck when trying to understand the energy and momentum of a photon, which has no mass? Not quite. You can divide the momentum by energy and have the mass drop out — and this works even for photons. So dividing the equation for momentum by the equation for energy, you get the following:

$$\frac{p}{E} = \frac{v}{c^2}$$

For photons, $v = c$ so

$$\frac{p}{E} = \frac{1}{c}$$

$$p = \frac{E}{c}$$

And for photons, $E = hf$, so

$$p = \frac{hf}{c}$$

And you may notice that $c = \lambda f$, so for a photon, the following is true:

$$p = \frac{h}{\lambda}$$

Putting all this together, Compton was able to show that you can relate the wavelength of the incident and scattered photons like this:

$$\lambda_{\text{scattered photon}} - \lambda_{\text{incident photon}} = \frac{h}{mc}(1 - \cos\theta)$$

where h is Planck's constant (6.626×10^{-34} joule-seconds), m is the mass of the electron (9.11×10^{-31} kilograms), and θ is the scattering angle of the photon, as Figure 13-3 shows earlier in this section.

So the difference in wavelength between the incident photon and the scattered photon varies from zero if the photon continues on its way undeflected ($\theta = 0°$) to h/mc if the photon is scattered through 90° ($\theta = 90°$). In fact, the quantity h/mc comes up frequently in Compton scattering, so it's called the *Compton wavelength:*

$$\lambda_{\text{Compton}} = \frac{h}{mc}$$

The Compton wavelength is equal to the following:

$$\lambda_{\text{Compton}} = 2.43 \times 10^{-12} \text{ m}$$

So using the Compton wavelength, the formula for Compton scattering gives you

$$\lambda_{\text{scattered photon}} - \lambda_{\text{incident photon}} = \lambda_{\text{Compton}} (1 - \cos \theta)$$

Compton scattering really put the issue to rest — photons can act as particles. The blackbody experiments (which I discuss earlier in "Blackbody Radiation: Discovering the Particle Nature of Light") gave rise to the idea that light was quantized, and Einstein explained the photoelectric effect by saying light came in energy packets (see the earlier section "Light Energy Packets: Advancing with the Photoelectric Effect"), but what really hit the ball over the wall was the Compton effect.

The de Broglie Wavelength: Observing the Wave Nature of Matter

In 1924, a physics grad student, Louis de Broglie, came up with an incredibly bold suggestion: He proposed that physicists radically change their ideas of the nature of particles without any direct experimental grounds for doing so. Physicists had already discovered the particle aspects of light waves, but there was no evidence that compelled physicists to drastically alter their ideas of particles.

However, de Broglie felt nature would be more beautiful if there was a kind of symmetry whereby particles could also behave as waves. Because photons obey the following equation (which you see in the preceding section):

$$p = \frac{h}{\lambda}$$

perhaps electrons and other particles would obey this equation:

$$\lambda = \frac{h}{p}$$

That is, perhaps particles of matter have a wavelength, and it's given by h/p. Amazingly, de Broglie turned out to be right. This section explains the experiments that supported this idea and then shows you how the math works.

Interfering electrons: Confirming de Broglie's hypothesis

Experiments have borne out de Broglie's idea. Early experiments were performed by bombarding nickel crystals with electrons and getting a diffraction pattern, just as you would from any wave. More recently, physicists sent electrons through a double-slit setup, producing the distinctive double-slit interference pattern (see Chapter 11 for info on light interference).

These physicists had a machine that emitted streams of electrons, and one day, they decided to pass the stream of electrons through a double-slit arrangement — the kind that gives rise to interference patterns with light waves. A funny thing happened on the way to the lecture hall: After adjusting the distance the double slits were apart, the *same* kind of interference pattern appeared — light and dark bars — on a photographic film (which records the positions at which the electrons strike it). The resulting light and dark bars looked exactly like an interference pattern, as Figure 13-4 shows.

Figure 13-4:
The inter-
ference
pattern of
electrons
sent through
dual slits.

That was an amazing result for the time — a stream of electrons passing through a double slit and creating an interference pattern, just like light. The electrons were acting like waves, so the world had to come to grips with this new idea that electrons could act as waves or particles.

What this means is that physicists needed to change their mental picture of electrons. No longer could one comfortably think of electrons as small pool balls, orbiting around the nucleus of an atom. Instead, physicists had to think in terms of tiny wave-like packets of matter.

Calculating wavelengths of matter

De Broglie stated that the wavelength of an electron (λ) equals Planck's constant ($h = 6.626 \times 10^{-34}$ joule-seconds) divided by momentum *(p)*:

$$\lambda = \frac{h}{p}$$

So do only electrons have a de Broglie wavelength? No, any object that has a momentum has a de Broglie wavelength, although the wavelength of objects you can see with the naked eye is vanishingly small. In this section, you calculate wavelengths of both electrons and larger objects.

Finding an electron's de Broglie wavelength

Try some numbers to see how de Broglie's wavelength works. For example, say you set an electron loose in your home and it starts zipping around at 1.9×10^6 meters per second. What's its de Broglie wavelength? The electron's speed is a nonrelativistic speed, far short of the speed of light in a vacuum, c, so the momentum of the electron is given by

$$p = mv$$

The mass of an electron is 9.11×10^{-31} kilograms, so the electron's momentum is

$$p = (9.11 \times 10^{-31} \text{ kg})(1.9 \times 10^6 \text{ m/s}) \approx 1.74 \times 10^{-24} \text{ kg-m/s}$$

So the electron's de Broglie wavelength is

$$\lambda = \frac{h}{p} = \frac{6.626 \times 10^{-34} \text{ J-s}}{1.74 \times 10^{-24} \text{ kg-m/s}} \approx 3.81 \times 10^{-10} \text{ m}$$

That means the electron's wavelength is 0.381 nanometers — about a thousand times smaller than the wavelength of visible light.

Finding the de Broglie wavelength of visible objects

Any object that has a momentum has a de Broglie wavelength. At about 0.381 nanometers, the de Broglie wavelength of the electron in the preceding section is huge compared to the de Broglie wavelength of an object visible to the naked eye.

Say that you're determined to see the de Broglie wavelength for yourself, and you've decided to throw a baseball past a wavelength meter. You throw the baseball, mass 0.150 kilograms, at a hefty 90.0 miles per hour. What's its de Broglie wavelength?

First, find out how fast the baseball is going in meters per second. As everybody knows, 1 meter per second is about 2.23693629 miles per hour. Okay, that's a little ridiculous on the significant digits, but you find that 90.0 miles per hour equals the following:

$$v = \frac{90.0 \text{ mph}}{1} \times \frac{1 \text{ m/s}}{2.24 \text{ mph}} \approx 40.2 \text{ m/s}$$

So the baseball's momentum is

$$p = mv = (0.150 \text{ kg})(40.2 \text{ m/s}) = 6.03 \text{ kg-m/s}$$

Now you can figure out the de Broglie wavelength of the baseball using the equation:

$$\lambda = \frac{h}{p}$$

$$\lambda = \frac{6.626 \times 10^{-34} \text{ J-s}}{6.03 \text{ kg-m/s}} \approx 1.10 \times 10^{-34} \text{ m}$$

So the wavelength of the baseball is 1.10×10^{-34} meters, or 1.10×10^{-25} nanometers — an incredibly small distance. You cannot measure distances that small. I realize saying that is risky in a field as unpredictable as physics, but this wavelength is on the order of the Planck length.

Some theories about quantum gravity say that *Planck length* is the length at which the apparently continuous structure of the universe breaks down. At such a minute scale, quantum physics dominates, and what used to be measurements can be discussed only in terms of probabilities. So how small is the Planck length? It would take 100,000,000,000,000,000,000 (that is, 10^{20}) Planck lengths to reach across a single proton. And that's about the wavelength of a baseball going 90 miles per hour. Blows the mind, doesn't it?

Not Too Sure about That: The Heisenberg Uncertainty Principle

You may have heard of the uncertainty principle — it's one of those physics concepts that has gravitated to everyday speech, as in, "Where's little Jimmy?" . . . "I don't know — the closer you try to pin him down, the farther away he'll be. You know, the uncertainty principle of children."

You take a look at the actual uncertainty principle here, including a derivation of the equation from what you see in the preceding section on matter waves.

Understanding uncertainty in electron diffraction

Figure 13-5 shows a stream of electrons going through a single slit and creating a single-slit diffraction pattern on a screen (see Chapter 11 for more on single-slit diffraction patterns). In the days of Newton, you wouldn't expect to see a diffraction pattern at all when you passed a stream of electrons through a single slit. You'd expect to see an exact image of the single slit on the screen (if you use photographic film as the screen, the pattern would be recorded on it).

Today, however, you know better. You know that you get a diffraction pattern — that is, a central bright bar surrounded by dark bars and lesser bright bars, as Figure 13-5 shows. Here's the insight this brings: When you're dealing with the small world (like electrons), you can no longer express things exactly.

For any individual electron going through the single slit, you can't say exactly where it's going to end up on the screen — it could end up anywhere there's a bright bar in the diffraction pattern. You can't assume that the electron will just keep going straight. You can speak of the electron's location on the screen only in terms of probabilities — and as you send more and more electrons through the slit, you'll end up with the diffraction pattern eventually.

Deriving the uncertainty relation

Say that the wavelength of the electrons passing through a single slit is λ and that the slit width is Δy, as in Figure 13-5. You can find the angle, θ, of the first dark bar in the diffraction pattern (as indicated in the figure) with the following equation:

$$\sin \theta = \frac{\lambda}{\Delta y}$$

In other words, θ tells you the angular width of the central bright bar (where the electron will land about 85 percent of the time). And if θ is small, $\sin \theta$ is about equal to $\tan \theta$ (that is, for small angles, $\sin \theta \approx \tan \theta$), so you have the following relation:

$$\tan \theta \approx \frac{\lambda}{\Delta y}$$

But what is the wavelength of the electron, λ? That's where de Broglie comes in, because you know that for matter waves, the following is true (from the earlier section "The de Broglie Wavelength: Observing the Wave Nature of Matter"):

$$\lambda = \frac{h}{p_x}$$

where p_x is the momentum of the electrons in the x direction and h is Planck's constant (6.626×10^{-34} joule-seconds).

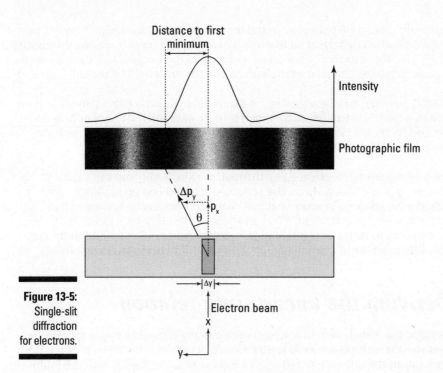

Figure 13-5:
Single-slit
diffraction
for electrons.

Substituting this value for λ into the equation for tan θ gives you this result:

$$\tan\theta \approx \frac{h}{p_x\Delta y}$$

So far, so good. Now take a look at Figure 13-5. If the electrons enter the slit with momentum p_x, then after going through the slit, they acquire an unknown momentum of Δp_y in the y direction (before the slit, you're assuming the electrons' momentum in the y direction was zero). Therefore, you have this relation between p_x and Δp_y:

$$\tan\theta = \frac{\Delta p_y}{p_x}$$

So setting the two equations for tan θ equal to each other gives you this result:

$$\frac{h}{p_x\Delta y} \approx \frac{\Delta p_y}{p_x}$$

Multiplying both sides by p_x and solving for h leaves you with

$$\frac{h}{\Delta y} \approx \Delta p_y$$

$$\Delta p_y \Delta y \approx h$$

And actually, this is pretty close to the real Heisenberg uncertainty principle, which says that

$$\Delta p_y \Delta y \geq \frac{h}{2\pi}$$

or in a more generic form:

$$\Delta p \Delta x \geq \frac{h}{2\pi}$$

Where Δp and Δx are the uncertainty in momentum and position respectively, the Heisenberg uncertainty relation says that the uncertainty in an object's momentum multiplied by the uncertainty in position must be greater than or equal to $h/2\pi$. In fact, $h/2\pi$ is so common that it got its own name, $\hbar$ (pronounced "h-bar"), so you often see Heisenberg's uncertainty relation written like this:

$$\Delta p \Delta x \geq \hbar$$

where

- $\hbar = \frac{h}{2\pi}$ (Planck's constant divided by 2π)
- Δp is the uncertainty in a particle's momentum
- Δx is the uncertainty in a particle's position

Here's how to think of the uncertainty relation for electrons passing through a single slit: By localizing the electrons to Δy (the slit), you introduce an uncertainty in the momentum Δp_y such that $\Delta p_y \Delta y \geq \hbar$.

As you can see from Heisenberg's relation, the more accurately you know a particle's momentum — that is, the smaller the uncertainty in momentum, Δp — the bigger the uncertainty in position, Δx. Conversely, the more accurately you know a particle's position — that is, the smaller the uncertainty in position, Δx — the bigger the uncertainty in momentum, Δp.

As a matter of fact, the Heisenberg uncertainty principle can also connect the energy of a particle, E, with the time that particle has that energy, t, like this:

$$\Delta E \Delta t \geq \hbar$$

where

- $\hbar = \dfrac{h}{2\pi}$ (Planck's constant divided by 2π)
- ΔE is the uncertainty in a particle's energy
- Δt is the uncertainty in the time interval during which the particle is in this state

Calculations: Seeing the uncertainty principle in action

The Heisenberg uncertainty principle shows an inverse relationship: The more accurately you know the position of a particle, the less accurately you can know its momentum, and vice versa.

In this section, you plug in some numbers to see how pinpointing one measurement leads to less accuracy in the other.

Finding uncertainty in speed, given an electron's position

Say that you use your new super-duper (and entirely theoretical) microscope to pin down the position of an electron to 1.00×10^{-11} meters. What's the minimum uncertainty in the electron's speed? Heisenberg tells you that

$$\Delta p \Delta x \geq \hbar$$

So the minimum uncertainty in the electron's momentum is

$$\Delta p_{min} = \frac{\hbar}{\Delta x}$$

Putting in the numbers gives you

$$\Delta p_{min} = \frac{6.626 \times 10^{-34}\ \text{J-s}}{2\pi\left(1.00 \times 10^{-11}\ \text{m}\right)} \approx 1.05 \times 10^{-23}\ \text{kg-m/s}$$

What's the uncertainty in speed? Well, for a nonrelativistic particle, $p = mv$, so

$$\Delta p = m\Delta v$$

$$\Delta v = \frac{\Delta p}{m}$$

Therefore, the minimum uncertainty on speed is

$$\Delta v_{min} = \frac{1.05 \times 10^{-23} \text{ kg-m/s}}{9.11 \times 10^{-31} \text{ kg}} \approx 1.15 \times 10^7 \text{ m/s}$$

So if you know an electron's position to within 1.0×10^{-11} meters of its actual location, you can only narrow its speed down to something within 1.15×10^7 meters per second of the actual speed — a mere 25,700,000 miles per hour.

That, of course, brings up the question of how you can measure anything about something moving at 25,700,000 miles per hour with respect to you, and the answer is that it'd be very difficult.

Finding uncertainty in position, given speed

Say that you want to hold an electron virtually still, at 1.00×10^{-5} meters per second — how closely can you localize it? At 1.00×10^{-5} meters per second, the electron's momentum is

$$\Delta p = m\Delta v = (9.11 \times 10^{-31} \text{ kg})(1.00 \times 10^{-5} \text{ m/s})$$

$$\approx 9.11 \times 10^{-36} \text{ kg-m/s}$$

And you can find the minimum uncertainty on position, given an uncertainty in momentum, like this:

$$\Delta x_{min} = \frac{\hbar}{\Delta p}$$

$$\Delta x_{min} = \frac{6.626 \times 10^{-34} \text{ J-s}}{2\pi \left(9.11 \times 10^{-36} \text{ kg-m/s} \right)} \approx 11.6 \text{ m}$$

So if you pin down the speed of an electron to within 1.0×10^{-5} meters per second of the actual speed, you can't locate it to less than 11.5 meters of its true location. Pretty slippery things, electrons!

Chapter 14

Getting the Little Picture: The Structure of Atoms

*W*ith the physics of the atom, you can understand an incredibly broad range of the behavior of your world. When you know that ordinary matter is composed of atoms, you can see why it comes in three main states: solid, liquid, and gas. You can understand much about the structure of solids, such as crystals, and the origin of laws of pressure in gasses. You can understand how atoms stick together to make molecules. You can picture much of how atoms and molecules interact with each other — in short, you can understand chemistry. That chemistry is vital to the comprehension of the functioning of living cells, so it's central to biology, too. If you want a basis for understanding so much of the world, knowing about the atom is definitely worth your while!

Although you can see atoms with various high-tech microscopes, atoms are invisible to the naked eye. Thus, it should come as no surprise that nobody knew how atoms were built until the early 20th century. Although Greek philosophers had theorized about atoms as the building blocks of matter, no one knew their structure — not Galileo, not Benjamin Franklin, not Isaac Newton — until the early 1900s.

The story of the atom — what makes it work and gives it the properties you observe today — is the subject of this chapter. You find out how physicists first came to accept a model that involved electrons orbiting around a nucleus. You also see refinements to that model and info on what quantum physics says about atomic structure.

Figuring Out the Atom: The Planetary Model

In the early 20th century, the accepted model of the atom was an English one. People already knew that the atom consisted of equal parts of positive charges and negative charges. J. J. Thomson, who discovered the electron, suggested that the positive charge was spread throughout the atom in a sort of "paste or pudding" material. The whole atom was filled with positive paste. The negative charges were embedded in the positive paste and held suspended there. That was the picture — negative charges suspended like plums in the positive paste. Physicists already knew that the negative charges were very light.

This model became known as the *plum pudding model* of the atom. And it was universally accepted until a physicist named Rutherford came up with an experiment to challenge it. This section describes that experiment and how it led to a planetary model of the atom.

Rutherford scattering: Finding the nucleus from ricocheting alpha particles

As in so many other experiments since the discovery of the atom, Ernest Rutherford (along with his students Hans Geiger and Ernest Marsden) needed a very tiny tool to probe very small distances. He chose a beam of charged particles moving at light speeds.

Some radioactive materials emit *alpha particles,* which have a double positive charge and are pretty massive (physicists now know that alpha particles are the nuclei of helium atoms, but in those days that was unknown, which is why they're called *alpha particles,* not *helium nuclei*). Rutherford directed a stream of alpha particles at a piece of gold foil and took a look at the results on screens that surrounded the foil target. The screens were arranged so that the widest possible range of angles of deflection could be observed (nearly the full 360° around the target). You can see his setup in Figure 14-1.

Classically, you'd expect all the alpha particles to go through the gold foil undeflected — alpha particles are relatively massive, and you couldn't expect the super-lightweight negative charges in the plum pudding model to deflect them. The positive charge was thought to be spread too thin to offer any resistance to the intruding alpha particles.

What happened instead was that many alpha particles were deflected as they passed through the foil, as you can see in the figure. In fact, some bounced off the target and reversed direction entirely. Rutherford said that it was "as incredible as if you fired a 15-inch shell at a piece of tissue paper and it came back and hit you."

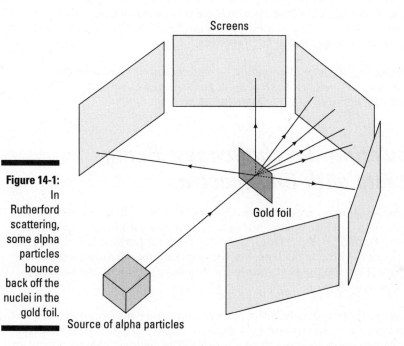

Figure 14-1:
In Rutherford scattering, some alpha particles bounce back off the nuclei in the gold foil.

Clearly, the classical plum pudding model needed some refinement. Rutherford proposed that the positive charge in an atom must be concentrated into a very small volume — the *nucleus*. That was the beginning of the modern theory of atomic structure.

Today, physicists know that although the atom is small — about 10^{-10} meters — the nucleus at the center is even smaller — about 10^{-15} meters. Put another way, if the nucleus was about the size of the width of a dime, the whole atom would be about a kilometer in diameter. So as you can see, the nucleus is tiny compared with the atom.

Collapsing atoms: Challenging Rutherford's planetary model

The tiny size of the nucleus led Rutherford to model the atom after the solar system. After all, the planets orbit around the sun, so the parallel was a natural one. The electrons were the planets, and the sun was the nucleus.

But as soon as this model was proposed, other physicists attacked it. If the electrons were simply in orbit around the nucleus, they would undergo a centripetal acceleration. And when electrons are accelerated, they radiate electromagnetic radiation — light (as I discuss in Chapter 7). So you'd expect the atom to radiate light, and then as the electrons lost energy, they'd collapse into the nucleus. In theory, an atom based on the simple planetary model would last only about 10^{-10} seconds.

That didn't square with observation — matter is pretty stable; it doesn't just collapse on itself in a flash of light. So it seemed clear that the simple planetary model needed some adjustment, too.

Answering the challenges: Being discrete with line spectra

The simple planetary model of the atom had problems because electrons would radiate and lose energy, eventually dropping into the nucleus. That problem puzzled physicists for quite some time. Rutherford had proven the existence of the atomic nucleus, but the planetary model didn't seem to work right, because the electrons in the atomic orbit would just radiate away their energy.

Or would they? In the early 20th century, atoms were the hot topic, and researchers did many experiments. Some of those experiments, which I discuss in this section, included taking a look at the electromagnetic spectrum that atoms emitted when they were heated.

Observing free atoms in gases

The electromagnetic spectrum emitted by a solid, such as a filament in a light bulb, is continuous. That is, a continuous range of wavelengths is emitted, some of which are in the visible range. And that led physicists to disregard the planetary model, because they assumed that electrons would just keep emitting radiation until they fell into the nucleus.

In a solid, atoms are strongly influenced by their neighbors. All the atoms bound in the solid emit and absorb light, and because they're all bound atoms, you end up with a continuous spectrum of wavelengths of light, some visible, some not. But when you take a look at free atoms in gases, not in solids, the story is a different one.

When you look at atoms in a heated gas, those atoms are free to do their own things, as individual atoms. That's when their individual atomic characteristics take over. The spectrum observed from heated gases turns out *not* to be continuous — the wavelengths are discrete. That is, only certain wavelengths are present.

For example, take a look at Figure 14-2, which shows the emitted spectrum of hydrogen in the visible region of the electromagnetic spectrum. Note that only specific wavelengths are present in the light spectrum emitted by hydrogen — which indicates that something then-unknown was going on with the atom.

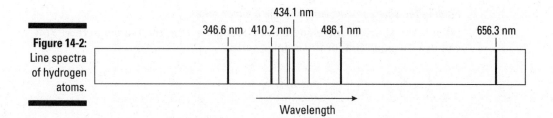

Figure 14-2:
Line spectra
of hydrogen
atoms.

Wavelength

The rejection of the planetary model of the atom was based on the assumption that electrons would just keep emitting, at ever-lower energies, until they fell into the nucleus. But experiment indicated that electrons weren't free to emit just any old wavelength — they had to emit only certain wavelengths. Perhaps there was hope for the planetary model after all.

Identifying wavelength patterns with the Lyman, Balmer, and Paschen series

Experimentally, hydrogen was observed to create a number of different series of lines with a definite pattern of wavelengths. The series repeated itself throughout the spectrum — in the infrared, visible, ultraviolet, and other parts of that spectrum.

Three such series are the Lyman, Balmer, and Paschen series, and researchers have observed that their particular patterns of wavelengths match. Although the actual wavelengths of each series are different, each series has the same characteristics — many close-together wavelengths, then wavelengths more spread apart, ending with some wavelengths very far apart.

Experimenters discovered equations that give you the wavelengths of these series. Here's how to get the wavelengths of these series of hydrogen:

- **Lyman series:** $\frac{1}{\lambda} = R\left(\frac{1}{1^2} - \frac{1}{n^2}\right)$ $n = 2, 3, 4, \ldots$

- **Balmer series (the visible series):** $\frac{1}{\lambda} = R\left(\frac{1}{2^2} - \frac{1}{n^2}\right)$ $n = 3, 4, 5, \ldots$

- **Paschen series:** $\frac{1}{\lambda} = R\left(\frac{1}{3^2} - \frac{1}{n^2}\right)$ $n = 4, 5, 6, \ldots$

In all these series, R is the *Rydberg constant,* 10,973,731.6 m^{-1}, or about 1.097×10^7 m^{-1}.

The discovery that atoms — in particular, hydrogen atoms — radiated light only at specific wavelengths gave life to the planetary model of the atom again, because it said that not just any wavelengths were allowed; instead, only particular wavelengths were allowed, which implied that electrons couldn't just keep radiating ever-smaller amounts of energy.

Finding the shortest wavelength in the Balmer series

What is the shortest wavelength in the visible series, the Balmer series? You can use the equation for the Balmer series:

$$\frac{1}{\lambda} = R\left(\frac{1}{2^2} - \frac{1}{n^2}\right) \qquad n = 3, 4, 5, \dots$$

The shortest wavelength (λ) corresponds to the biggest reciprocal ($1/\lambda$), and you get that when the $1/n^2$ term goes to zero:

$$\frac{1}{n^2} \to 0$$

And that means that $n \to \infty$. So for the shortest wavelength of the Balmer series, you have the following:

$$\frac{1}{\lambda} = R\frac{1}{2^2}$$
$$\frac{1}{\lambda} = \frac{1.097 \times 10^7 \, \text{m}^{-1}}{2^2}$$
$$\frac{1}{\lambda} \approx 2.74 \times 10^6 \, \text{m}^{-1}$$

Taking the reciprocal gives you the shortest wavelength in the Balmer series. Here's what you get:

$$\lambda \approx \frac{1}{2.74 \times 10^6 \, \text{m}^{-1}} \approx 3.65 \times 10^{-7} \, \text{m} = 365 \, \text{nm}$$

So that's the shortest wavelength in the Balmer series: 365 nm, which is deep violet.

Finding the longest wavelength in the Balmer series

How about that longest wavelength on the Balmer series? You can use the equation for the Balmer series:

$$\frac{1}{\lambda} = R\left(\frac{1}{2^2} - \frac{1}{n^2}\right) \qquad n = 3, 4, 5, \dots$$

The longest wavelength (λ) corresponds to the smallest reciprocal ($1/\lambda$), and you get that when the $1/n^2$ term is as large as possible — that is, when $n = 3$

(that's the minimum possible value for the n in the Balmer series). So for the longest wavelength of the Balmer series, you have

$$\frac{1}{\lambda} = R\left(\frac{1}{2^2} - \frac{1}{3^2}\right)$$

$$\frac{1}{\lambda} \approx 1.52 \times 10^6 \, \text{m}^{-1}$$

Taking the reciprocal gives you the longest wavelength in the Balmer series:

$$\lambda = \frac{1}{1.52 \times 10^6 \, \text{m}^{-1}} \approx 6.56 \times 10^{-7} \, \text{m} = 656 \text{ nm}$$

And that's the longest wavelength in the Balmer series: 656 nm, which corresponds to red. So the Balmer series neatly encompasses most of the visible spectrum — deep violet to red.

Fixing the Planetary Model of the Hydrogen Atom: The Bohr Model

A physicist named Niels Bohr came up with a new model of the hydrogen atom. He merged the new quantum ideas of Max Planck and Albert Einstein to come up with the Bohr model of the atom. He postulated that electrons were allowed only certain energies in atoms — they couldn't have just any energy. That idea was close to the theories on quantization in blackbody radiation, which I cover in Chapter 13.

Bohr's model of the atom violated the law of electromagnetism, which said that accelerating charges radiate electromagnetic waves. But Bohr didn't care much; the current laws of physics couldn't explain how the atom worked (at least, not insofar as predicting the line spectra like the Balmer series), and his could.

Bohr arranged the electrons in an atom in specific *orbits,* which corresponded to the allowed total-energy levels for the electrons in terms of kinetic and electric potential energy. You can see the Bohr model illustrated in Figure 14-3. Note how this model is consistent with the findings of Rutherford: that the positive charges should be concentrated together rather than smoothly distributed (like the plum pudding model).

Furthermore, Bohr took Einstein's then-new ideas on photons and said that when electrons go from a higher orbit (with more energy and a larger orbital radius) to a lower one (with less energy and a smaller orbital radius), a photon is emitted by the electron. The fact that only specific orbits are allowed accounts for the specific wavelengths in the spectrum of gases (for details, see the earlier section "Answering the challenges: Being discrete with line spectra").

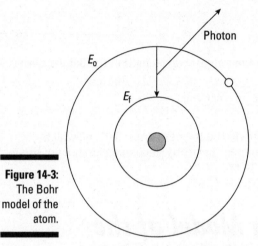

In other words, when an electron falls from a higher orbit, where the electron has total energy E_o, to a lower one, where the electron has energy E_f, the energy of the emitted photon, hf, is

$$hf = E_o - E_f$$

where h is Planck's constant, 6.626×10^{-34} J-s, and f is the frequency of the photon. The Bohr model did explain the observed atomic spectra well and so gradually came to be accepted.

Finding the allowed energies of electrons in the Bohr atom

After developing his model, Bohr turned to trying to calculate the allowed energy levels for the electrons in the atom. An electron's total energy is the sum of its kinetic and potential energy:

$$E = KE + PE$$

And that looks like this:

$$E = \frac{1}{2}mv^2 - \frac{kZe^2}{r}$$

where m is the mass of the electron, v is its speed, k is Coulomb's constant, Z is the number of protons in the nucleus, e is the electron's charge, and r is the radius of the electron's orbit (see Chapter 3 for more on the potential energy of point charges).

How do you figure out mv^2? The centripetal force on an electron is equal to

$$F = \frac{kZe^2}{r^2}$$

And centripetal force also equals

$$F = \frac{mv^2}{r}$$

So you can set the two forces equal and solve for mv^2:

$$\frac{mv^2}{r} = \frac{kZe^2}{r^2}$$
$$mv^2 = \frac{kZe^2}{r}$$

Substituting this into the equation for total energy gives you

$$E = \frac{1}{2}\frac{kZe^2}{r} - \frac{kZe^2}{r}$$
$$E = \frac{-kZe^2}{2r}$$

And that's the expression for the energy of an electron in an atom. *Note:* It's negative because the electron is bound in the atom.

Getting the allowed radii of electron orbits in the Bohr atom

Bohr's equation for the energy of an electron in an atom doesn't do you much good until you figure out the allowed values of the radius r.

Bohr proposed that angular momentum — that is, the angular momentum of electrons — was quantized in atoms. Angular momentum, L, equals

$$L = mvr$$

So Bohr imposed this quantization on angular momentum:

$$L_n = mv_n r_n = \frac{nh}{2\pi} \qquad n = 1, 2, 3, \ldots$$

where h is Planck's constant, 6.626×10^{-34} J-s, and n is an integer, where the various allowed values of the angular momentum correspond the values of n. In other words, he said that angular momentum is a multiple of $h/2\pi$.

Solving this equation for v_n gives you

$$v_n = \frac{nh}{2\pi m r_n} \qquad n = 1, 2, 3, \ldots$$

In the preceding section, you see that

$$mv^2 = \frac{kZe^2}{r}$$

Finding the speed squared gives you

$$v_n^2 = \frac{n^2 h^2}{4\pi^2 m^2 r_n^2} \qquad n = 1, 2, 3, \ldots$$

And mv_n^2 is

$$mv_n^2 = \frac{n^2 h^2}{4\pi^2 m r_n^2} \qquad n = 1, 2, 3, \ldots$$

So set your two values of mv^2 equal to each other:

$$\frac{kZe^2}{r_n} = \frac{n^2 h^2}{4\pi^2 m r_n^2} \qquad n = 1, 2, 3, \ldots$$

Solving for r_n here gives you

$$r_n = \frac{h^2}{4\pi^2 m k e^2} \frac{n^2}{Z} \qquad n = 1, 2, 3, \ldots$$

And that's what the allowed Bohr radii are. With this information, you can finally calculate the allowed energies.

Here are the allowed Bohr radii in meters:

$$r_n = \left(5.29 \times 10^{-11} \text{m}\right) \frac{n^2}{Z} \qquad n = 1, 2, 3, \ldots$$

Finding allowed energy for hydrogen using the Bohr radius

Now go back and find the allowed energy levels. You know that

$$E = \frac{-kZe^2}{2r}$$

So substituting in for r gives you

$$E_n = \frac{-2\pi^2 mk^2 e^4}{h^2} \frac{Z^2}{n^2} \qquad n = 1, 2, 3, \ldots$$

Here's what that looks like in joules:

$$E_n = \left(-2.16 \times 10^{-18}\right)\left(\frac{Z^2}{n^2}\right) \qquad n = 1, 2, 3, \ldots$$

And here's what the allowed energy levels look like in electron volts:

$$E_n = \left(-13.6\right)\left(\frac{Z^2}{n^2}\right) \qquad n = 1, 2, 3, \ldots$$

For example, what's the lowest energy an electron can have in hydrogen? That would be when $n = 1$ and Z (the number of protons in the atom) = 1, so you have the following:

$$E_1 = \left(-13.6\right)\left(\frac{1^2}{1^2}\right) \qquad \text{Hydrogen, } n = 1$$
$$= -13.6 \text{ eV}$$

So that's the most tightly you can bind an electron in hydrogen: −13.6 electron-volts. That is, it takes 13.6 electron-volts of energy to free an electron in the $n = 1$ state, also called the *ground state*.

How about the $n = 2$ state of hydrogen? The energy of this level looks like this:

$$E_2 = \left(-13.6\right)\left(\frac{1^2}{2^2}\right) \qquad \text{Hydrogen, } n = 2$$
$$= -3.4 \text{ eV}$$

So the energy with which an electron is bound in the $n = 2$ state of hydrogen is −3.4 electron-volts. As you can see, the energy levels of the successively higher states decrease, until you ultimately get a bound state energy of 0 — which means the electron isn't bound at all.

Finding allowed energy levels for lithium ions, Li²⁺

Here's something you should know: The Bohr model applies only to atoms with one electron. That's because it's still a relatively simple model that doesn't take into account the interaction between electrons (all electrons repel each other, and that affects their total energy).

So if you want to use the Bohr model for, say, lithium, which has three protons in its nucleus ($Z = 3$), you can use the equation for the energy levels only if you have doubly ionized lithium — that is, if you have a lithium atom where two of the electrons have been removed and you have only one electron left (such lithium ions have a net positive charge of +2). For doubly ionized lithium, the energy of the ground state is

$$E_1 = (-13.6)\left(\frac{3^2}{1^2}\right) \qquad \text{Lithium, } n = 1$$
$$\approx -122 \text{ eV}$$

So the ground state of doubly ionized lithium has an energy of 122 electron-volts.

Finding the Rydberg constant using the line spectrum of hydrogen

With Bohr's picture of the hydrogen atom in mind, the general equation for the wavelength of electron transition in hydrogen is

$$\frac{1}{\lambda} = R\left(\frac{1}{n_f^2} - \frac{1}{n_i^2}\right) \qquad n_f = 2, 3, 4, \ldots \qquad n_i = n_f + 1,\, n_f + 2,\, n_f + 3,\, \ldots$$

where n_i is the initial energy level of the electron and n_f is the final energy level of the electron.

In the earlier section "Identifying wavelength patterns with the Lyman, Balmer, and Paschen series," the wavelengths of the Balmer series of hydrogen is given by

$$\frac{1}{\lambda} = R\left(\frac{1}{2^2} - \frac{1}{n^2}\right) \qquad n = 2, 3, 4, \ldots$$

That's because the Balmer series includes electron transitions from higher orbits to the $n = 2$ state.

The Lyman series is just the series with $n_f = 1$:

$$\frac{1}{\lambda} = R\left(\frac{1}{1^2} - \frac{1}{n_i^2}\right) \qquad n_i = 2, 3, 4, \ldots$$

And the Paschen series is the series with $n_f = 3$:

$$\frac{1}{\lambda} = R\left(\frac{1}{3^2} - \frac{1}{n_i^2}\right) \qquad n_i = 4, 5, 6, \ldots$$

Furthermore, the energy levels of an electron in hydrogen are

$$E_n = \frac{-2\pi^2 mk^2 e^4}{h^2} \frac{1}{n^2} \qquad n = 1, 2, 3, \ldots$$

For photons, $E = hf$. And because $f = \frac{c}{\lambda}$, that means that

$$E = \frac{hc}{\lambda}$$

$$\frac{1}{\lambda} = \frac{E}{hc}$$

So substitute the value of E_n into the equation for $1/\lambda$, and it becomes

$$\frac{1}{\lambda} = \frac{2\pi^2 mk^2 e^4}{h^3 c}\left(\frac{1}{n_f^2} - \frac{1}{n^2}\right) \qquad \begin{array}{l} n_f = 2, 3, 4, \ldots \\ n_i = n_f + 1, n_f + 2, n_f + 3, \ldots \end{array}$$

In other words, the Bohr theory predicts that the Rydberg constant, R, is equal to

$$R = \frac{2\pi^2 mk^2 e^4}{h^3 c} \quad \ldots$$

Wow, that's quite a lot of letters. And it turns out to be completely right. So Bohr's model of the hydrogen atom predicts the Rydberg constant as a combination of other constants. Cool!

Putting it all together with energy level diagrams

Keeping track of the electron transitions in an atom like hydrogen can be tough. You have the $3 \rightarrow 2$ transition (that is, from the third excited state to the second excited state), the $5 \rightarrow 3$ transition, the $7 \rightarrow 5$ transition, and, of course, many more.

To help keep track of all the possible transitions, you can look at energy level diagrams. You can see an example for hydrogen in Figure 14-4. There, the various energy levels are represented as horizontal lines (some of the lines are marked with their energy levels in electron-volts in the figure).

Transitions from one energy level to another appear as downward-pointing arrows. Thus, you can see that the first line in the Lyman series goes from the –3.4 electron-volt energy level to the –13.6 energy level (which means that the emitted photon has energy of 13.6 – 3.4 eV = 10.2 eV), and so on.

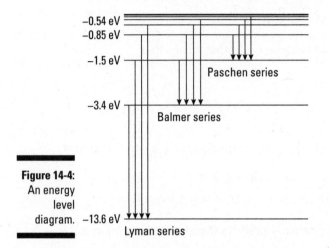

Figure 14-4:
An energy
level
diagram.

You can see the various popular series of the hydrogen atom marked according to their energy level transitions in the figure (many more such series in addition to those shown exist).

De Broglie weighs in on Bohr: Giving a reason for quantization

Louis de Broglie (who shows up in the discussion of matter waves in Chapter 13) took a look at Bohr's theory and asked why angular momentum should be quantized like this:

$$L_n = mv_n r_n = \frac{nh}{2\pi} \qquad n = 1, 2, 3, \ldots$$

In other words, why are only specific angular momentums allowed?

De Broglie's explanation was that if you think of the electron in motion around the nucleus in terms of matter waves (which were his specialty), then the quantization condition simply becomes the following: A full wavelength (or two wavelengths, or some integer number) must equal the circumference of the electron's orbit, $2\pi r$, where r is the radius of the electron's orbit.

In other words, according to de Broglie, the quantization of angular momentum really meant that

$$2\pi r = n\lambda \qquad n = 1,\ 2,\ 3,\ \dots$$

And because, according to de Broglie's theory, the wavelength of a matter wave is $\lambda = \dfrac{h}{p}$, you can substitute this in for λ to give you

$$2\pi r = n\frac{h}{p} \qquad n = 1,\ 2,\ 3,\ \dots$$

And because $p = mv$, you have

$$2\pi r = n\frac{h}{mv} \qquad n = 1,\ 2,\ 3,\ \dots$$

So solving for mvr, the angular momentum, gives you

$$mvr = n\frac{h}{2\pi} \qquad n = 1,\ 2,\ 3,\ \dots$$

But because the angular momentum, L, equals mvr, this is simply the Bohr quantization of angular momentum:

$$L_n = mv_n r_n = n\frac{h}{2\pi} \qquad n = 1,\ 2,\ 3,\ \dots$$

And that's a very nice result. Instead of saying that the angular momentum must be quantized — it's not obvious why this should be so — you instead say that a multiple of the electron's matter wave must be equal to the circumference of each orbit. Chalk another one up for de Broglie.

Electron Configuration: Relating Quantum Physics and the Atom

Bohr's model of the hydrogen atom was a good one, and it was a milestone. For the first time, the allowed energy levels of an atom were not allowed to be continuous but were found to be discrete — that is, only specific allowed radii of the electrons were permitted: $n = 1, 2, 3$, and so on.

Because only specific values are allowed, n became the first *quantum number* of the atom. The *quantum number* tells you which of the allowed states that a particle (such as an electron) is in. This section introduces four quantum numbers and explains their significance.

Understanding four quantum numbers

Physics has so far found four quantum numbers for the electrons in an atom. I explain all four of them here.

The principal quantum number, n

The *principal quantum number* is n, the quantum number that Bohr discovered, corresponding to which orbit the electron is in. The ground state has $n = 1$, the next level has $n = 2$, and so on. Each successive energy level corresponds to an orbital radius farther away from the nucleus.

The orbital angular momentum quantum number, l

In addition to the principal quantum number, electrons also have an *angular momentum quantum number,* which is given the letter l. This quantum number indexes which of the permitted angular-momentum states that an electron is in.

The l quantum number can vary from 0 to $n - 1$. The total angular momentum, L, of an electron with angular momentum quantum number l is

$$L = \left(l(l+1)\right)^{1/2}\frac{h}{2\pi} \qquad l = 0, 1, 2, 3, ..., n-1$$

When the angular momentum quantum number of an electron is zero, the electron has no net angular momentum.

The quantum-mechanical picture of the atom does not see the electron as a particle orbiting the nucleus. Instead, the electron is a kind of wave that's allowed to have certain specific configurations. So you don't talk of electron orbits but electron *states*, which correspond to the different configurations of the waves. The wave quantifies the probability of finding the electron at any given point. If the magnitude of the wave is larger, then the electron is more likely to be found there if you measure its position.

You can see a picture of this wave for an electron in a zero angular-momentum state ($l = 0$) in Figure 14-5. The electron wave is a simple sphere for the zero angular-momentum state.

Figure 14-5:
Electrons in an $l = 0$ state.

The electron wave can look decidedly strange if the electron has a nonzero angular momentum quantum number. For example, take a look at the orbit of an $l = 2$ electron in Figure 14-6. As l increases (up to $n - 1$), the orbits of electrons can become correspondingly complex.

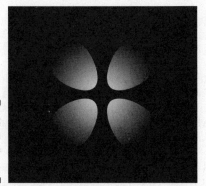

Figure 14-6:
Electrons in an $l = 2$, m = ±1 state.

The magnetic quantum number, m

The quantum number m is called the *magnetic quantum number* because it results when you apply a magnetic field to the atom. Applying a magnetic field to an electron that has angular momentum means that electron can have a component of angular momentum in the direction of the applied field. That direction of the magnetic field defines an axis, which is conventionally called the z axis in physics (as opposed to x or y).

The *magnetic angular momentum* is the component of the electron's angular momentum along the z-axis, so m can take values from $-l$ to $+l$ — which is why the magnetic quantum number is sometimes referred to as m_l. And because it represents the state of angular momentum in the z direction, it's also referred to as m_z in some physics books. But in this book, I simply use m.

The z component of the electron's angular momentum, L_z, is equal to the following:

$$L_z = m\frac{h}{2\pi} \qquad m = -l,\ -l+1,\ ...,\ l-1,\ l$$

Note that the magnetic quantum number, m, of the electron wave in Figure 14-6 is ± 1.

The spin quantum number, m_s (or s)

Finally, physicists discovered that each electron also has an intrinsic spin. That is, even when an electron is in an $l = 0$ state, where m is also 0, the electron still has an intrinsic spin. Electron *spin* something like the spin of the Earth, which spins about its axis, even as it rotates around the Earth.

The spin quantum number is given the letter s. Physicists have since determined that many subatomic particles, including photons, have an intrinsic spin. The spin of a photon is 1, and the spin of an electron is $s = \frac{1}{2}$.

Like the orbital angular momentum, spin can have a component along an applied magnetic field — m_s. The m_s quantum number is the fourth quantum number of an electron in an atom (the s quantum number is intrinsic to the electron and doesn't change, whether or not it's in an atom). For an electron, m_s can have the values $-\frac{1}{2}$ or $\frac{1}{2}$.

For an electron, $m_s = \frac{1}{2}$ is called *spin up* and $m_s = -\frac{1}{2}$ is called *spin down*, corresponding to the component of the electron's spin along or against an applied external magnetic field.

Number crunching: Figuring out the number of quantum states

Electrons in an atom can have four quantum numbers — n, l, m, and m_s. If you know the value of n, you can figure out the number of different quantum states an electron can have. Here's how it works:

1. **Use *n* to find the number of *l* states.**

   Suppose you want to find how many different quantum states an electron in the *n* = 2 energy level can have. The equation for *L* in the earlier section "Understanding four quantum numbers" tells you that *l* = 0, 1, 2, 3, ..., *n* – 1. So for *n* = 2, you can have angular-momentum quantum numbers *l* from 0 to *n* – 1, or 0 to 1. So that's two quantum states to start.

2. **Use the *l* states to find the number of *m* states.**

   The equation for the *z* component of the electron's angular momentum, L_z, in "Understanding four quantum numbers" says that *m* = –*l*, –*l* + 1, ..., *l* – 1, *l*. The *l* = 0 state can only have *m* = 0, but the *l* = 1 state can have *m* = –1, 0, or 1. So that's a total of four quantum states so far.

3. **Account for spin for each of the *m*, *l*, and *n* states.**

   The electron can also have a spin quantum number, m_s. No matter what the other quantum states of the electron, it can have either spin down or spin up — that is, m_s = –½ or m_s = ½. Thus the *n*, *l*, *m* state of 2, 0, 0 splits into two states:

   - **Spin down:** 2, 0, 0, –½

   - **Spin up:** 2, 0, 0, ½

   The 2, 1, 1 state also splits into two states:

   - **Spin down:** 2, 1, 1, –½

   - **Spin up:** 2, 1, 1, ½

   And the 2, 1, 0 and 2, 1, –1 states likewise split into two states. Each of the *n*, *l*, *m* states splits into two states when you add the spin of the electron: *n*, *l*, *m*, –½ and *n*, *l*, *m*, ½, so that converts the four quantum states into eight.

And that's the answer: An electron with atomic orbit number 2 can have a total of eight quantum states.

Figure 14-7 shows a tree diagram of the quantum numbers of the first few electron states in a hydrogen atom. In this figure, you can see how each quantum number introduces a new branch to the tree so that the number of possible states increases. (I explain the meaning of the labels on the far right of this figure later in "Using shorthand notation for electron configuration.")

Although there are four quantum numbers for every electron, only the principal quantum number, *n*, and the angular momentum quantum number, *l*, determine the energy level of the electron. The energy level is mostly determined by the principal quantum number, *n*, but there are some small differences in energy depending on the angular-momentum quantum number, *l*. Under these circumstances, the energy of the electron state does not depend on the other two quantum numbers. However, if a magnetic field is present, then there can be an interaction that causes the other two quantum numbers to change the energy level of the state.

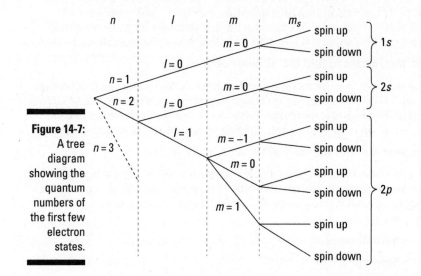

Multi-electron atoms: Placing electrons with the Pauli exclusion principle

Not every atom in the world is a hydrogen atom (thank goodness). Other atoms have more electrons, and the electrons can interact with each other.

In hydrogen, you have only one proton, so the *atomic number, Z,* equals one. In neutral atoms, Z always equals the number of electrons in an atom (because it's the same as the number of protons). The one electron in hydrogen spends most of its time in the ground state, $n = 1$. If it's in a higher principal quantum level, it emits a photon and drops down to the ground state pretty quickly — it doesn't stay in the excited states for long.

So what's preventing all the electrons in an atom with a higher Z value from all dropping down to the ground state? For example, say you have a nickel atom, which has $Z = 28$. The electrons in an atom of nickel are in all kinds of states: $n - 1, n - 2, n - 3$, and so on. Why don't they all just fall down to the $n = 1$ ground state?

Bonding with atoms: The physics of chemistry

The structure of particular atoms and their electron states means that they like to combine with other atoms in very particular ways. So when atoms and molecules get together, they can reconfigure or react. The study of how atoms and molecules react is exactly what you study in chemistry.

For instance, atoms often bond to each other to create molecules. Chemical bonds usually involve the atoms' outermost electrons *(valence electrons)* because they're in higher energy states, so it takes less energy to remove them. Hence, these electrons are least tightly bound to the atom.

Ionic bonds occur when an electron in one atom is in a particular energy state and there's an unoccupied state that has a lower energy in a nearby atom. The atoms can then exchange the electron, leaving the atom that the electron came from positively charged and the atom that it moved to negatively charged — the atoms become *ionized*. The two charged atoms then stick together by the electrostatic force between them, forming an *ionic bond*.

At other times when atoms come together, the shape of one of the electron waves becomes distorted until it's spread between the two atoms. Two electrons with opposite spins can sit in this state. When this happens, these electrons are effectively shared between the two atoms, resulting in a *covalent bond*.

Atomic structure also plays a role in the release of heat during some chemical reactions. A simple example of a chemical reaction is the burning of coal. There's a very strong covalent bond between oxygen atoms and carbon atoms because of the configurations of their electron states. When these atoms get close enough, they can combine very violently because of the strength of the bond. This violent motion causes the neighboring atoms to vibrate — this is the heat released in the burning of the carbon in oxygen.

The *Pauli exclusion principle* says that no two electrons in any atom can have the same combination of all four quantum numbers. This means you can only have two electrons in the ground state of any atom, because $n = 0$ only allows $l = 0$, $m = 0$, and spin up ($\frac{1}{2}$) or spin down ($-\frac{1}{2}$) electrons. The $n = 2$ state can only have eight electrons (as you find out in the preceding section) and so on.

At first, Wolfgang Pauli proposed his exclusion principle simply to describe the structure of the atom as deduced in experiments. Without it, there was no reason all the electrons didn't just collapse into the same, lowest energy state. Later on, as quantum mechanics developed, this principle found a strong theoretical footing as it applied to all particles with non-integer spin, not just the electron.

Say you're building a nickel atom from scratch, particle by particle. You assemble the protons and neutrons into the atom's nucleus first. Then you start adding electrons:

1. **The first two electrons go into the ground state, the $n = 1$ state.**

   This state corresponds to a particular radius from the nucleus where the electron wave has greatest magnitude; it's sometimes called the *K shell.*

2. **Next, start stocking up the $n = 2$ state; eight electrons total would fit into that state.**

   This state is sometimes called the *L shell.*

3. **Put electrons in the next level, the $n = 3$ level.**

   The *M shell* holds up to 18 electrons. Along with the 2 electrons in the *K* shell and the 8 in the *L* shell, this accounts for all 28 electrons.

With all these quantum numbers available, it can be pretty tough to keep track of. So in time, physicists developed a shorthand notation, which I explain next.

Using shorthand notation for electron configuration

Take a piece of boron, atomic number 5. What's the outermost electron in any one of the atoms? Why, that's the $2p^1$ electron. That's the kind of shorthand that scientists developed for naming the state of an electron. Each number, letter, and superscript number represents something:

- **The first number is the principal quantum number.** For the outermost electron in boron, the 2 in $2p^1$ means the principal quantum number, *n*, is 2.

- **The letter stands for the angular-momentum quantum number, *l*.** And here's where it gets tricky. Historically, the different values of *l* have been assigned different letters, and here they are:

    - $l = 0 \rightarrow s$
    - $l = 1 \rightarrow p$
    - $l = 2 \rightarrow d$
    - $l = 3 \rightarrow f$

- $l = 4 \rightarrow g$
- $l = 5 \rightarrow h$

and so on. So in $2p^1$, the p stands for an angular-momentum quantum number of 1.

✔ **The superscript indicates the electron's quantum number for the angular momentum.** Thus, the first electron you add to an atom is the $1s^1$ electron, and the next is the $1s^2$ electron.

The next electron is the $2s^1$ electron, followed by the $2s^2$ electron. Then come the $2p^1$, $2p^2$, $2p^3$, and so on electrons.

So in terms of the full electron configuration, you can denote a boron atom ($Z = 5$) as $1s^2 2s^2 2p^1$, where each electron is in the lowest available energy state.

To better see how the shorthand relates to the quantum numbers, check out Figure 14-7, earlier in this chapter. The labels at the right show you how two electrons can fit in the $1s$ subshell (where $n = 1$, $l = 0$), two electrons fit in the $2s$ subshell (where $n = 2$, $l = 0$), and six electrons fit in the $2p$ subshell (where $n = 2$, $l = 1$). If you were to draw more of the chart, you could see that because of the way the tree diagram branches off, two electrons fit in s subshells, six fit in p subshells, ten fit in d subshells, fourteen fit in f subshells, and so on.

Table 14-1 shows the states of the electrons for atoms with values of Z up to 18. Remember that these electron configurations are for an atom in its lowest energy state. For example, the hydrogen atom has a single electron in the 1s state when it's at its lowest energy. However, if the atom is excited, the electron would be in a higher shell and have non-zero angular momentum (quantum number l).

Table 14-1   Electron Configurations for the First 18 Elements

Element	*Z*	*Electron Configuration*
Hydrogen	1	$1s$
Helium	2	$1s^2$
Lithium	3	$1s^2 2s$
Beryllium	4	$1s^2 2s^2$
Boron	5	$1s^2 2s^2 2p$
Carbon	6	$1s^2 2s^2 2p^2$
Nitrogen	7	$1s^2 2s^2 2p^3$

(continued)

Table 14-1 *(continued)*

Element	Z	Electron Configuration
Oxygen	8	$1s^2 2s^2 2p^4$
Fluorine	9	$1s^2 2s^2 2p^5$
Neon	10	$1s^2 2s^2 2p^6$
Sodium	11	$1s^2 2s^2 2p^6 3s$
Magnesium	12	$1s^2 2s^2 2p^6 3s^2$
Aluminum	13	$1s^2 2s^2 2p^6 3s^2 3p$
Silicon	14	$1s^2 2s^2 2p^6 3s^2 3p^2$
Phosphorous	15	$1s^2 2s^2 2p^6 3s^2 3p^3$
Sulfur	16	$1s^2 2s^2 2p^6 3s^2 3p^4$
Chlorine	17	$1s^2 2s^2 2p^6 3s^2 3p^5$
Argon	18	$1s^2 2s^2 2p^6 3s^2 3p^6$

Chapter 15

Nuclear Physics and Radioactivity

- -

In This Chapter

▶ Understanding nuclear structure

▶ Looking at the force holding protons and neutrons together

▶ Understanding alpha, beta, and gamma decay

▶ Measuring radioactivity

- -

An atom's electron structure (which I cover in Chapter 14) is what gives an atom its chemical properties. Elements act chemically depending on the outermost shell of electrons. But the electrons are only part of the story. You also have the nucleus, which is the subject of this chapter. The electrons in an atom orbit around the relatively small but dense nucleus, and the nucleus makes up by far the most mass in an atom.

Although you don't deal with the nucleus in general chemistry classes, you certainly do in physics. Physicists can probe the nucleus using subatomic particles, and as a result, people know a great deal about the nucleus. And of course, the nucleus is where the radioactivity of atoms is centered. So in this chapter, you explore the structure of the nucleus, examine the forces that hold protons and neutrons together, and find out what happens when other forces prevail and the atom undergoes radioactive decay.

Grooving on Nuclear Structure

The nucleus sits at the center of the atom. At one time, people thought it was completely solid, with all the positive charge in the atom concentrated in it. The nucleus was thought of as a tiny sphere, on the order of 10^{-15} meters.

How small is 10^{-15} meters? Well, put another way, how big is 10^{15} meters? That's about 10,000 times the distance to the sun, and 1 meter is to 10^{15} meters what 10^{-15} meters is to 1 meter.

Scientists now know that picture is wrong; the nucleus has a great deal of structure. It's made up of various nucleons, as Figure 15-1 shows. No doubt you've heard of the two types of *nucleons* — protons and neutrons:

- ✔ **Protons:** Protons are tiny, positively charged particles of a miniscule mass, about 1.672×10^{-27} kilograms. Although that makes them sound small, they're huge compared with the true lightweight of the atom, the electron, with a mass of 9.11×10^{-31} kilograms (so the proton is about 1,800 times more massive than the electron). The charge of a proton is 1.60×10^{-19} coulombs, exactly the same magnitude (although opposite in sign) as the electron's charge.

- ✔ **Neutrons:** Neutrons are electrically neutral particles that are more massive than electrons — and slightly more massive than protons. Neutrons have a mass of around 1.675×10^{-27} kilograms, compared with the proton's mass of 1.672×10^{-27} kilograms.

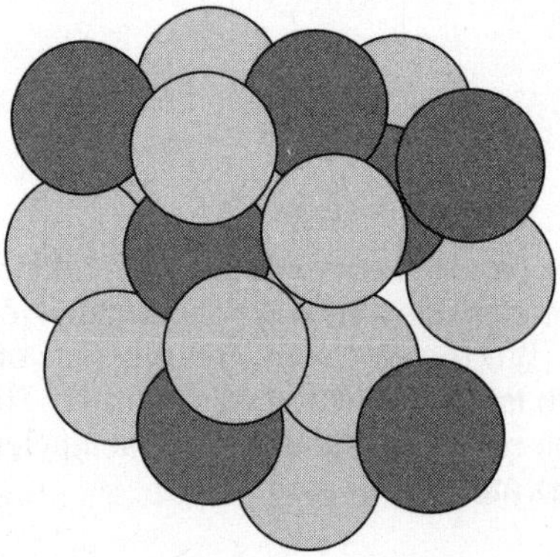

Figure 15-1:
An atomic
nucleus.

Note that neutrons are more massive than protons. In a way, you can almost think of neutrons as combinations of protons and electrons, which results in a neutral particle. Although that picture isn't exact, neutrons can decay — and when they do, they produce a proton and an electron.

So is the nucleus just a bundle of nucleons (neutrons and protons)? Are nuclei just spherical packs of nucleons? Pretty much. Experiments show that the nucleus has a roughly spherical structure and that it's indeed made up of bundles of separate nucleons. So the image in Figure 15-1 is actually a pretty accurate one.

Now for a little chemistry: Sorting out atomic mass and number

The number of protons in an atom set the atomic number, Z, which tells you what kind of atom you're dealing with. For example, if $Z = 2$, you have a helium atom. If $Z = 6$, you have a carbon atom. So that's the main connection between chemistry and the nucleus: The atomic number of the atom determines which element you have.

The number of neutrons in an atom, given the letter N, doesn't affect most chemical processes. Chemically speaking, neutrons are inert.

Taken together, the number of neutrons and protons, N and Z, is the *atomic mass number* (or *nucleon number*), A:

$$A = N + Z$$

So is the mass of an atom just A multiplied by the average mass of a nucleon (the average of a proton mass and a neutron mass)? Approximately, yes. I say *approximately* because some nuclei have unequal numbers of protons and neutrons, as I show you in the next section.

In terms of chemistry and the periodic table, you use a particular shorthand when indicating elements, like this: $^{12}_{6}C$. That's the symbol for standard carbon (C), which has six protons ($Z = 6$) and atomic mass number 12 ($A = 12$).

In general, the symbol used for an element is

$$^{A}_{Z}X$$

where A is the element's atomic mass number, Z is the atomic number, and X is the one- or two-letter symbol for the element (like H for hydrogen, He for helium, C for carbon, and of course, Os for osmium, as everyone knows).

Neutron numbers: Introducing isotopes

The atomic number of an atom determines the element you're dealing with. So $Z = 6$ is carbon. But the number of neutrons, which don't affect the chemical properties of an atom for the most part, can actually vary. Thus, you have two forms of carbon appearing in nature. The first form has 6 protons, of course, and 12 total nucleons (protons plus neutrons):

$$^{12}_{6}C$$

But some uppity carbon atoms — about 1.10 percent — have 13 nucleons, not just 12, so their symbol is

$$^{13}_{6}C$$

TECHNICAL STUFF

Weighing out a set number of atoms with moles

When the number of molecules or atoms is important, scientists can pull out another quantity: the *mole*. You have 1 mole of a substance when you have a number of atoms or molecules that's equal to *Avogadro's number*, which equals about 6.022×10^{23}.

Like the atomic mass unit, the mole is defined in terms of the carbon-12 atom: 1 mole of a substance is the quantity that has the same number of atoms (or molecules if it's a molecular substance) as the number of atoms in 12 grams of carbon-12. When you have a mole, you have 6.022×10^{23} atoms (or molecules).

To get a mole of another element, find the element's atomic mass number and write the units as grams instead of atomic mass units. Then weigh it out.

These two forms of carbon are called *isotopes* of carbon, atoms of the same element that differ in the number of neutrons.

REMEMBER

Isotopes are often denoted with their atomic mass numbers, so the first carbon isotope, $^{12}_{6}\text{C}$, is carbon-12, and the second, $^{13}_{6}\text{C}$, is carbon-13. You may also see them denoted as C-12 and C-13 or even as C12 and C13.

So why do you see a symbol like this in the periodic table of the elements?

$$^{12.011}_{6}\text{C}$$

That's because the atomic mass number in the periodic table, 12.011, is the *average* atomic mass number of all carbon atoms that occur naturally.

REMEMBER

The unit of measurement for atomic mass is the aptly named *atomic mass unit (amu)*, which equals 1.66×10^{-27} kilograms. Hence, the average carbon nucleus has a mass of

$$^{12.011}_{6}\text{C mass} = (12.011)\left(1.66 \times 10^{-27} \text{ kg}\right) \approx 1.99 \times 10^{-26} \text{ kg}$$

The atomic mass unit is technically defined to be one-twelfth of the mass of a carbon-12 atom (that is, six protons, six neutrons, and twelve electrons composing the atom). But this unit is a convenient scale to use for all atoms.

You may notice that an amu has slightly less mass than either nucleon, the proton or neutron, and wonder why. That's because the mass of a nucleus has slightly less mass than the sum of the masses of its individual nucleons. You can understand this strange result because some mass goes into the binding energy to hold the nucleus together (a topic that's coming up in the section "Hold on tight: Finding the binding energy of the nucleus").

Not all isotopes are equally stable. As you see later in "Understanding Types of Radioactivity, from α to γ," nuclei can undergo radioactive decay and alter the number of nucleons in the nucleus.

Boy, that's small: Finding the radius and volume of the nucleus

Experiments have shown that the radius of the nucleus is about equal to the following, where A is the number of nucleons (protons and neutrons):

$$r \approx (1.2 \times 10^{-15} \text{ m})A^{1/3}$$

For example, what's the radius of the nucleus of the carbon-12 atom, the most common form of carbon? Plugging numbers in the formula gives you the following:

$$r \approx (1.2 \times 10^{-15} \text{ m})(12)^{1/3} \approx 2.7 \times 10^{-15} \text{ m}$$

So how small is that? Comparing the radius of a carbon-12 nucleus to 1 meter is like comparing the thickness of a dime to the distance between Earth and Saturn. In other words, the nucleus is pretty small.

The nucleus is a roughly spherical bundle of nucleons, so its volume is roughly equal to the volume of a sphere. The volume is approximately given by

$$V \approx \frac{4}{3}\pi r^3$$

Substituting in 1.2×10^{-15} meters $\times A^{1/3}$ — the value of r that was experimentally determined — you get this approximate expression for the volume of an atom's nucleus:

$$V \approx \frac{4}{3}\pi\left(1.2 \times 10^{-15}\text{m}\right)^3 A$$

Calculating the density of the nucleus

What's the density, ρ, of the nucleus? Well, density equals mass divided by volume:

$$\rho = \frac{m}{V}$$

The mass of a carbon-12 atom is 12 amu (atomic mass units). An amu is about 1.66×10^{-27} kilograms, so

$$_6^{12}\text{C mass} = (12)\left(1.66 \times 10^{-27} \text{ kg}\right) \approx 1.99 \times 10^{-26} \text{ kg}$$

You can figure out the volume of the carbon-12 nucleus as roughly

$$V \approx \frac{4}{3}\pi r^3$$

where $r = 2.7 \times 10^{-15}$ meters (as you calculate in the preceding section). Therefore, you can say that

$$V \approx \frac{4}{3}\pi \left(2.7 \times 10^{-15} \text{m}\right)^3 \approx 8.2 \times 10^{-44} \text{m}^3$$

Therefore, the density is

$$\rho \approx \frac{m}{V} = \frac{1.99 \times 10^{-26} \text{ kg}}{8.2 \times 10^{-44} \text{m}^3} \approx 2.4 \times 10^{17} \text{ kg/m}^3$$

And that's dense — a pea-sized piece of pure nuclear material would weigh about 27,000,000,000 metric tons.

The Strong Nuclear Force: Keeping Nuclei Pretty Stable

If you think about it, nuclei shouldn't hold together at all. After all, a nucleus can contain dozens of protons, as well as neutrons, which means that there are dozens of strong positive charges very, very, very close to each other. And you know what happens when positive charges get close to each other: They repel each other. At very close range, that repelling force can get to be huge. Because the size of the nucleus is about 10^{-15} meters, the outward force on the protons in a nucleus is enormous.

So why don't nuclei fly apart immediately? Why don't they explode? This section explains the forces at work.

Finding the repelling force between protons

You can calculate the electrostatic force exerted by two protons on each other in a nucleus with this equation (from Chapter 3):

$$F = \frac{kq_1q_2}{r^2}$$

where k is the constant 8.99×10^9 N $\cdot$ m^2/C^2, q_1 and q_2 are the two charges, and r is the distance between the charges.

Plugging in the numbers and assuming the distance between protons is 10^{-15} meters gives you

$$F = \frac{\left(8.99 \times 10^9 \, \text{N} \cdot \text{m}^2/\text{C}^2\right)\left(1.6 \times 10^{-19}\text{C}\right)\left(1.6 \times 10^{-19}\text{C}\right)}{\left(1.0 \times 10^{-15}\text{m}\right)^2} \approx 230 \text{ N}$$

So that's 230 newtons — about 52 pounds! That's an incredible force between two protons, so why on Earth don't they fly apart?

Holding it together with the strong force

The protons in a nucleus don't fly apart because as strong as the electrostatic force is, the *strong nuclear force* is even stronger. The strong force works between nucleons, and it's what keeps the nucleus together.

The strong force is one of only four fundamental forces discovered (so far) in nature. You know all about two of those forces already — the gravitational force and the electrostatic force. The other two are the strong force and the weak force. I discuss the strong force in this section.

The limits of the strong nuclear force

The *strong force* binds nucleons in the nucleus together, and it's constantly fighting against the electrostatic force. So if the strong force is so strong, why doesn't it take over everything — why doesn't everything just collapse into one gigantic nucleus? That's just the thing: The strong force is effective only over very small distances — about 10^{-15} meters. Beyond that range, it's effectively zero. It's almost as if its express purpose is to hold nuclei together and nothing more.

Here's where it gets interesting: The strong force pulls nuclei together (there's no repelling strong force), and the electrostatic force pushes them apart. There comes a time when you get so many protons together in the nucleus that their mutually repelling force starts to overcome the strong force. This happens because the strong force has such a limited range — nucleons have to be essentially right next to each other to be bound by the strong force. But the electrostatic force is long-range, so while one proton is bound to two or three other nucleons by the strong force, as you add more protons, they all gang up on the first proton with their electrostatic forces.

There comes a time when the electrostatic repulsion between protons overcomes the strong force holding them in place and bingo — the nucleus explodes. And that's exactly where radioactivity comes from: The electrostatic repulsion of many protons overcomes the attraction of the strong force holding them together.

The stabilizing power of neutrons

What about neutrons? They don't repel other nucleons, and they can exert the strong attractive force, so shouldn't it follow that the more neutrons you have in a nucleus, the more stable it is? And that's what happens for the most part: As the atomic number (the number of protons) increases, you need more and more neutrons to keep the nucleus stable. Adding more neutrons helps separate the volatile protons while exerting a stabilizing strong force.

So as nuclei get bigger and bigger — as you add more and more protons — you need to add even more neutrons to keep things stable. The biggest atom that's generally considered stable is $^{209}_{83}$Bi. That's bismuth (yes, the stuff that makes antidiarrhea medicines work), with 83 protons. It takes 126 neutrons to hold things in place.

Because of the details of how the strong force works, protons and neutrons like to pair up. If you get too many neutrons, then there's a kind of imbalance in the energies between the neutrons and protons, and the nucleus becomes unstable. So nuclei with roughly equal numbers of protons and neutrons are the most stable — though because of the repulsion between the protons, larger nuclei need relatively more neutrons. The result is quite a narrow range of stable combinations of neutrons and protons, where their numbers are roughly equal, but the relative number of neutrons increases slightly as the total number of nucleons increases.

As you go higher in atomic number (to higher values of Z), you don't get enough neutrons to hold the nucleus together forever. For example, uranium, which is famously radioactive, has $Z = 92$. So now you know where radioactivity comes from. Pretty cool, eh?

Hold on tight: Finding the binding energy of the nucleus

The strong force is what holds nuclei together — which means that separating the nucleons in a nucleus takes work. That work is called the *binding energy* of the nucleus.

How do you find the binding energy of a nucleus without taking it all apart? You can be clever about this and measure a nucleus's mass compared to the masses of its constituent nucleons. That is, when you find the mass of the nucleus, the nucleus is less massive than the sum of the nucleons that go into it (so you could say that a nucleus is less than the sum of its parts). Why? Because some mass went into the binding energy of the nucleus.

The difference between the masses of all the nucleons separately and the final nucleus is called the *mass defect* of the nucleus, which has the symbol Δm. So the mass defect of a nucleus is

$$\Delta m = \sum m_{\text{nucleons}} - m_{\text{nucleus}}$$

where $\sum m_{\text{nucleons}}$ is the sum of the masses of the nucleons and m_{nucleus} is the mass of the nucleus after all the nucleons are put together.

So how can you find the binding energy from the mass defect? You may recall that Einstein said that $E_0 = mc^2$, so the binding energy of a nucleus is equal to

$$E_{\text{binding}} = \Delta mc^2$$

where Δm is the mass defect of the nucleus.

Finding the mass defect

Check out some numbers. For example, grab a standard helium atom: ^4_2He. The nucleus of that atom has two protons and two neutrons, and experiments show that it has a mass of 6.6447×10^{-27} kilograms. What is its mass defect?

You can find the mass defect of a nucleus with $\Delta m = \sum m_{\text{nucleons}} - m_{\text{nucleus}}$. Here, that becomes

$$\Delta m_{\text{He-4}} = 2m_{\text{proton}} + 2m_{\text{neutron}} - m_{\text{nucleus}}$$

Like any good physicist, you know that

✔ The mass of a proton is 1.6726×10^{-27} kilograms

✔ The mass of a neutron is 1.6749×10^{-27} kilograms

So you have the following:

$$\Delta m_{\text{He-4}} = 2(1.6726 \times 10^{-27} \text{ kg}) + 2(1.6749 \times 10^{-27} \text{ kg}) - (6.6447 \times 10^{-27} \text{ kg})$$

$$\approx 0.0503 \times 10^{-27} \text{ kg} = 5.03 \times 10^{-29} \text{ kg}$$

That's very small indeed.

Calculating the binding energy

What's the binding energy of the standard helium atom? That equals $E_{\text{binding}} = \Delta mc^2$. The mass defect, m, is 5.03×10^{-29} kilograms, and the speed of light is approximately 3.00×10^8 meters per second, so you have the following:

$$E_{\text{binding}} = (5.03 \times 10^{-29} \text{ kg})(3.00 \times 10^8 \text{ m/s})^2$$

$$\approx 4.53 \times 10^{-12} \text{ J}$$

What's that in electron-volts, eV? An electron-volt is the energy needed to push one electron through 1 volt of electric potential, and 1 electron-volt is 1.60×10^{-19} joules, so the binding energy of He-4 is

$$E_{\text{binding}} = \left(4.53 \times 10^{-12} \text{ J}\right)\frac{1 \text{ eV}}{1.60 \times 10^{-19} \text{ J}} \approx 2.83 \times 10^7 \text{ eV}$$

So the binding energy is 28.3 million electron-volts. And because the proton has the same charge as the electron, 28.3 million electron-volts is the energy you get by letting a proton drop through 28.3 million volts.

Put another way, it takes 24.6 eV to pull an electron away from an atom of He-4. It takes more than 1 million times that amount to pull a proton out of the He-4 nucleus. That's the strong force at work — and it has to overcome the repulsive force of two protons at extremely close distance as well.

Understanding Types of Radioactivity, from α to γ

Radioactivity happens when atomic nuclei explode, and as you know, radioactivity can have some nasty side effects, such as radiation poisoning. But where others prudently refused to tread, physicists eagerly jumped in.

Radioactivity is the process by which unstable nuclei disintegrate. Such nuclei don't go quietly — they emit nuclear fragments and other various particles. In addition to fragments of nuclei, physicists recorded three types of particles emitted by radioactive elements:

✔ α (alpha) particles

✔ β (beta) particles

✔ γ (gamma) particles

As you may expect (because these are the first three letters of the Greek alphabet), these particles were named in the order of their discovery. They're all particles created by nuclear decay and so are fit topics for your study.

Physicists know how to handle such particles — with a magnetic field, which bends the trajectory of such particles and lets researchers determine more about the particles' mass and charge. Figure 15-2 shows such a setup. A radioactive source, enclosed in a lead container, is placed at the bottom of the apparatus. Nuclear decay byproducts — some charged, some not — shoot out of the container and, passing through a magnetic field, then hit a screen or detector to be recorded.

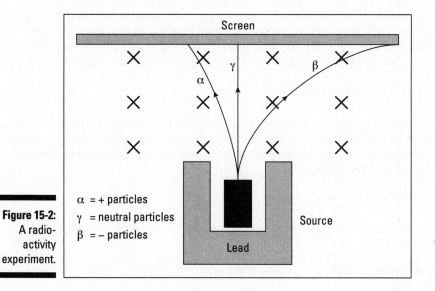

Figure 15-2:
A radio-
activity
experiment.

In their quest to understand nuclear reactions, nuclear physicists had these tools at their disposal:

✔ Conservation of total energy

✔ Conservation of charge

✔ Conservation of linear momentum

✔ Conservation of angular momentum

✔ Conservation of nuclear number

In this section, I introduce you to nuclear decay for all three types of particles — alpha, beta, and gamma — and explain how some of these laws of conservation come into play.

Releasing helium: Radioactive alpha decay

When you're talking about radioactivity, it's hard not to talk about uranium, which is what comes to most people's minds when the topic first pops up. Physicists found that a uranium-238 atom — U-238 for short — decays to thorium and an alpha particle. The nuclear reaction is denoted this way (it's something like the notation for chemical reactions):

$$^{238}_{92}U \rightarrow \, ^{234}_{90}Th + \alpha$$

Experiments like the one in Figure 5-2 eventually determined what the alpha particle is — it's a helium nucleus! (Notice how the atomic mass number decreased by 4 and the atomic number, the number of protons, went down by 2.) So $\alpha = \, ^4_2He$. Therefore, the U-238 decay becomes the following:

$$^{238}_{92}U \rightarrow \, ^{234}_{90}Th + \, ^4_2He$$

How much energy goes into the kinetic energy of the products of this decay (that is, the thorium atom and the alpha decay)? The kinetic energy would be the difference between the mass of U-238 and the byproducts of the decay, Th-234 and the alpha particle:

$$KE = m_{U\text{-}238} - m_{Th\text{-}234} - m_{He\text{-}4}$$

Looking up the mass of these atoms in atomic mass units, you get the following:

- ✔ **Uranium-238:** $m_{U\text{-}238} = 238.0508$ amu
- ✔ **Thorium-234:** $m_{Th\text{-}234} = 234.036$ amu
- ✔ **Helium-4:** $m_{He\text{-}4} = 4.0026$ amu

So the kinetic energy equal to the mass defect released is

mass defect = 238.0508 amu − 234.0436 amu − 4.0026 amu = 0.0046 amu

A mass of 1 amu, from $E_0 = mc^2$, is equivalent to an energy of 931.5 million electron-volts (mega electron-volts, or MeV), so the kinetic energy released is

$$KE = (0.0046 \text{ amu})(931.5 \text{ MeV}) \approx 4.3 \text{ MeV}$$

That kinetic energy is divided between the thorium atom and the alpha particle.

Charge and angular momentum are conserved as always. Note that linear momentum is also conserved, and because the thorium atom is about 60 times the mass of the alpha particle, the alpha particle ends up with about 60 times the speed of the thorium atom. Thus, alpha particles are the most prominent result of a lump of decaying U-238.

Another conservation law is in action in nuclear reactions: the conservation of the total number of nucleons. Note how in U-238 decay, uranium has 238 nucleons, and it decays into thorium, with 234 nucleons, along with the alpha particle with 4 nucleons. The sum of the nucleons on each side of the reaction is the same. You can use this to check whether the reaction is possible or to find out whether something is missing from your reaction.

Mass is not conserved in U-238 decay, because the end product has less mass than the original uranium atom. However, total energy is conserved, so if you take into account $E_0 = mc^2$, where mass and energy are equivalent, then mass-energy is conserved.

Gaining protons: Radioactive beta decay

Besides the alpha particle, another particle that's produced by radioactive decay is the beta particle. For example, thorium (which you see in uranium decay in the preceding section) can decay itself. In particular, the $^{234}_{90}\text{Th}$ isotope of thorium decays to $^{234}_{91}\text{Pa}$ (Pa is protactinium, a metallic element). This decay also produces a beta particle (β), so the reaction is written like this:

$$^{234}_{90}\text{Th} \rightarrow {}^{234}_{91}\text{Pa} + \beta$$

What on Earth is a β particle? After long experimentation, physicists determined it was an electron (or in some radioactive decays, the electron's positively charged antimatter counterpart, the *positron*).

So how do you denote an electron in format like $^{234}_{90}\text{Th}$? An electron contains no neutrons and certainly no protons. In fact, its charge is opposite to a proton, so perhaps you should designate it as $^{0}_{-1}\text{e}$. And that's exactly what physicists did, so the decay of $^{234}_{90}\text{Th}$ looks like this:

$$^{234}_{90}\text{Th} \rightarrow {}^{234}_{91}\text{Pa} + {}^{0}_{-1}\text{e}$$

Notice what's happening in beta decay. The net atomic number — that is, the number of protons, *Z,* actually *increases* as a result of the decay. How does that happen? It happens because a neutron decays into a proton and an electron (the beta particle). It's rare to have neutrons decay, but it does happen. Also note that when the electron is written as $_{-1}^{0}e$, then the sum of the upper and lower indexes on both sides of the reaction are the same.

Emitting photons: Radioactive gamma decay

Just like the whole atom can be in an excited state, the nucleus of an atom can exist in excited states. That is, just as electrons can be in higher orbits and jump down to lower orbits, emitting a photon (as I explain in Chapter 14), so can nucleons.

Does that mean that a nucleus can emit a photon, just like orbital electrons? Yes, it certainly can. For example, radium (Ra) nuclei can fire off a photon. You start with the excited form of the radium nucleus, which is denoted with an asterisk (*):

$$_{88}^{226}\text{Ra}^{*}$$

Then you decay to an unexcited form of the radium atom, along with a photon, which is a *gamma ray* (that is, a very high-power photon), which is denoted with the Greek letter γ like this:

$$_{88}^{226}\text{Ra}^{*} \rightarrow {}_{88}^{226}\text{Ra} + \gamma$$

And there you have it — excited radium emitting a high-power photon and turning into normal radium.

When nuclei emit high-energy photons, it's called *gamma decay.* Note that in gamma decay, the atomic number of the nucleus doesn't change — no charge is carried away by an emitted particle. Instead, a photon shoots out of the nucleus with a lot of energy and at high frequency.

How high is the frequency? Well, take a look. Say that you have the radium gamma decay $_{88}^{226}\text{Ra}^{*} \rightarrow {}_{88}^{226}\text{Ra} + \gamma$. The emitted photon has an energy of 0.186 mega electron-volts. For a photon, the energy is

$$E = hf$$

where h is Planck's constant and f is the frequency of the photon. So solving for f, you have

$$f = \frac{E}{h}$$

To plug in E, you need to find the energy of the photon (gamma ray) in joules. You know that the photon's energy is 0.186 mega electron-volts, or 1.86×10^5 electron-volts. There are 1.60×10^{-19} joules in 1 electron-volt, so do the conversion:

$$E = \left(\frac{1.86 \times 10^5 \, eV}{1} \right) \left(\frac{1.60 \times 10^{-19} \, J}{1 \, eV} \right) \approx 2.98 \times 10^{-14} \, J$$

So you have 2.98×10^{-14} joules to work with. That corresponds to a photon frequency of

$$f = \frac{E}{h} = \frac{2.98 \times 10^{-14} \, J}{6.63 \times 10^{-34} \, J\text{-s}} \approx 4.49 \times 10^{19} \, s^{-1}$$

So the frequency is about $4.49 \times 10^{19} \, s^{-1}$. What's the wavelength of the gamma ray? To find the wavelength, λ, you can use the relation $c = \lambda f$, or

$$\lambda = \frac{c}{f}$$

So you have

$$\lambda = \frac{c}{f} = \frac{3.00 \times 10^8 \, m/s}{4.49 \times 10^{19} \, s^{-1}} \approx 6.68 \times 10^{-12} \, m$$

In other words, the wavelength is about 6.68×10^{-3} nanometers. And that puts it at about one hundred-thousandth of the wavelength of visible light.

Grab Your Geiger Counter: Half-Life and Radioactive Decay

A radioactive element is one with an unstable nucleus. The nucleus decays into a more stable one and, in the process, produces a byproduct, such as an electron or photon (that is, a beta particle or gamma ray). You can detect this byproduct with a clever device called a *Geiger counter*. With a Geiger counter, you can take a single atom of a radioactive element and sit and wait to detect when it decays.

However, physicists found that you can't predict when a radioactive atom will decay — it's random. But because some elements decay more quickly than others, you can say that they're likely to decay sooner rather than later.

When you have a great number of unstable atoms together, then there's a definite average rate of decay. If you put your Geiger counter to the atoms, you detect a certain number of decays per second. This rate falls with time, because there are fewer and fewer unstable nuclei left to decay.

Say you go to the corner store and come back with a pound of radium, $^{226}_{88}\text{Ra}$ (don't try this at home — radium caused a lot of radiation poisoning in early physicists). Unwrapping your radium, you note that it's decaying slowly into radon, Rn, like this:

$$^{226}_{88}\text{Ra} \rightarrow \, ^{222}_{86}\text{Rn} + \alpha$$

Hmm, you think — radon is a gas. How long will this sample of radium last? To figure that out, you need to understand the concept of half-life, a convenient way to discuss the rate of decay. In this section, you look at the concepts of half-life and radioactivity.

Halftime: Introducing half-life

The rate at which you see radioactive nuclei decaying is in proportion to the number of atoms you have. This means that a radioactive substance decays exponentially. Exponential decay reduces the amount of radioactive substance by a constant fraction in equal time intervals.

When working with exponential decay problems, a convenient time interval to use is the time that the sample takes to reduce by half. *Half-life* tells you how long it takes for *half* of a given number of atoms to decay. (Any other fraction would work, but half is nice and simple, and it immediately gives you a good feel for how fast your sample is decaying.)

The half-life of $^{226}_{88}\text{Ra}$ is about 1,600 years, so in 1,600 years half your sample will have decayed into $^{226}_{80}\text{Rn}$, so you'll have half of your $^{226}_{88}\text{Ra}$ left in your sample. In another 1,600 years you'll be left with $(\frac{1}{2})^2$ of the original amount of $^{226}_{88}\text{Ra}$ in your sample. So as you can see, you don't have much to worry about — the radium in your sample will be around for quite some time.

What are the radioactive half-lives of various isotopes? Table 15-1 gives you a sampling in case you're interested.

Table 15-1	Half-Lives of Radioactive Substances	
Element	**Isotope**	**Half-Life**
Polonium	$^{214}_{84}Po$	1.64×10^{-4} seconds
Krypton	$^{89}_{36}Kr$	3.15 minutes
Radon	$^{222}_{86}Rn$	3.83 days
Strontium	$^{90}_{38}Sr$	29 years
Radium	$^{226}_{88}Ra$	1.6×10^3 years
Carbon	$^{14}_{6}C$	5.73×10^3 years
Uranium	$^{238}_{92}U$	4.47×10^9 years

For example, $^{226}_{86}Rn$, radon, is a radioactive gas created when radium decays. It turns out that radon can collect in the basements of houses, and it's a health worry because it's radioactive.

Say that you have a radon test done in your house that finds some radon gas — an estimated 100,000,000 (or 1.0×10^8) atoms. If you seal the house against more radon getting in, how many atoms are left after 31 days?

Here's how to find out how much of a sample is left using half-lives:

1. **To find out how many half-lives have passed, divide the amount of time that's passed by the length of a half-life.**

   The half-life of radon is 3.83 days, so 31 days is equal to about eight half-lives:

   $$\frac{31 \text{ days}}{3.83 \text{ days}} \approx 8.1 \text{ half-lives}$$

2. **Multiply ½ by itself, once for each half-life, to find the fraction of the sample that's left.**

   You can write the number of half-lives as an exponent on ½. That means that the radon sample will have decayed until

   $$\left(\frac{1}{2}\right)^8 = \frac{1}{2^8} = \frac{1}{256}$$

   of the original sample is left.

3. To get the number of radioactive atoms remaining, multiply the fraction from Step 2 by the number of atoms you started with.

Only $\frac{1}{256}$ of the original sample remains, or

$$\frac{1}{256}\left(1.0\times10^8 \text{ atoms}\right) = \frac{1.0\times10^8 \text{ atoms}}{256} \approx 3.9\times10^5 \text{ atoms}$$

So after a month, about 390,000 atoms will be left.

Decay rates: Introducing activity

How do you quantify the number of decays per second from some radioactive sample? You use the *activity* of the sample, which is given as the number of decays per second:

$$\text{Activity} = -\frac{\Delta N}{\Delta t}$$

where ΔN is the change in the number of radioactive nuclei in the time Δt. The negative sign indicates that the number of radioactive nuclei falls with time, so ΔN is negative (making the activity positive). Defined like this, the activity simply measures the rate at which the radioactive nuclei are decaying.

Radioactivity is measured in *becquerels* (Bq) — 1 becquerel equals one decay per second. You can also measure activities in *curies* (Ci), where

1 Ci = 3.70 × 10^{10} Bq

The rate at which the nuclei are decaying is in proportion to the number of nuclei you have, so the activity is also equal to

$$\text{Activity} = -\frac{\Delta N}{\Delta t} = \lambda N$$

where λ is called the *decay constant.*

The decay constant makes it easy to figure out how much of a sample you have left after a certain time with this equation:

$$N = N_0 e^{-\lambda t}$$

where N is the number of atoms you have currently, N_0 is the number of atoms you started out with, λ is the decay constant, and t is the time.

You can relate half-life, $T_{1/2}$, to the decay constant like this:

$$T_{1/2} = \frac{\ln(2)}{\lambda}$$

where *ln* means natural log (log to the base *e*). This also means that

$$\lambda = \frac{\ln(2)}{T_{1/2}}$$

For example, take a radon gas sample, $^{226}_{86}$Rn. What is its activity to start and after 31 days? To start, say you have $N = 1.0 \times 10^8$ atoms, and the half-life of radon is 3.83 days, which is 3.31×10^5 seconds. First find the decay constant:

$$\lambda = \frac{\ln(2)}{T_{1/2}} = \frac{0.693}{3.31 \times 10^5 \, \text{s}} \approx 2.09 \times 10^{-6} \, \text{s}^{-1}$$

Then use the radioactivity equation. The initial activity equals the following:

Activity = λN

$$= (2.09 \times 10^{-6} \, \text{s}^{-1})(1.0 \times 10^8 \, \text{atoms})$$

$$= 209 \, \text{s}^{-1}$$

So the initial activity is 209 becquerels.

After 31 days, about 8.1 half-lives, that number is cut down by a factor of 256 (as you find out in the preceding section); this means the final activity is $209 \div 256 \approx 0.82$ becquerels. Quite a change!

Part V
The Part of Tens

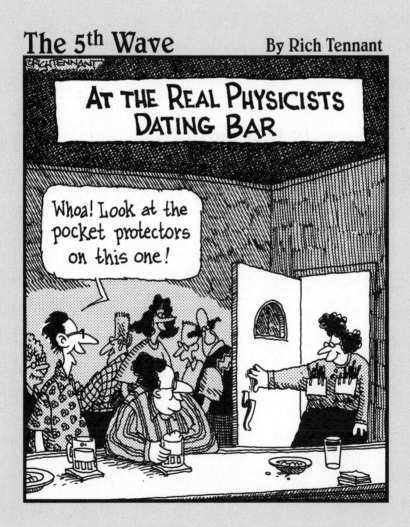

AT THE REAL PHYSICISTS DATING BAR

Whoa! Look at the pocket protectors on this one!

In this part . . .

Here, you see a rundown of ten physics experiments that changed the world. From measuring the speed of light to discovering radiation, it's all right here. I also give you a chapter covering the best online tools for problem-solving in physics.

Chapter 16

Ten Physics Experiments That Changed the World

*p*hysics has a way of changing the world, and this chapter considers ten physics experiments that did just that. Okay, so "changed the world" is a pretty bold statement, but it's true. You can see the impact of these discoveries in a number of ways. Taking the high-minded approach, you can appreciate how they revealed more of the universe's astonishing beauty. If you're a little more down-to-earth, you can observe how they've changed the way people think about the world and its possibilities.

And if you're really grounded, you can consider the technological advances born of the ten experiments in this chapter. From the treatment of cancer with Marie-Curie's radiation to night-vision goggles from the photoelectric effect, the practical applications have been numerous, and they're still increasing. In fact, the possible uses for quantum mechanics are just beginning to develop — sci-fi technology like teleportation and quantum computing may yet become more than a dream.

Regardless of how you look at these experiments, they were big in the world of physics. So go ahead — pull up a lab stool and read on. (And if you're inspired to try some radiation experiments of your own, don't forget your lead shielding.)

Michelson's Measurement of the Speed of Light

In the 19th century, people already knew that light was fast, but no one knew precisely how fast. In 1878, physics instructor Albert Abraham Michelson thought up an experiment to determine the speed of light. His setup dramatically improved upon previous estimates and signaled the beginning of his career — a pretty impressive one, at that. His work accompanied the transition from classical to modern physics.

Michelson placed a mirror far away from his setup and then devised a rotating eight-sided mirror that he sent a beam of light toward. The light bounced off one of the mirror's eight sides, sped to the far-away mirror, came back, and hit another side of the eight-sided mirror to go into a detector.

By making the rotating mirror (going at about 256 revolutions per second) synch with the arrival of the light from the far mirror, he was able to measure very short time intervals. So Michelson came up with a measurement of the speed of light: His value was 299,944 kilometers per second, plus or minus 51 kilometers per second. Current estimates put the speed of light at about 299,792 kilometers per second. Not bad, eh? You can read more on the speed of light in Chapter 8.

Young's Double-Slit Experiment: Light Is a Wave

The nature of light was a mystery in the early 19th century. No one really understood what it was — was it similar to something in other parts of nature, or was it something special, all its own?

Light has wave-like (as well as particle-like) qualities. Thomas Young did the first experiments to make the wave-like nature of light clear more than 200 years ago. In 1803, Young performed his famous double-slit experiment, which showed that light rays could interfere with other light rays in much the way ripples in a pond could interfere with other ripples. His paper, "Experiments and Calculations Relative to Physical Optics," became world famous. You can read about light wave interference in Chapter 11.

Jumping Electrons: The Photoelectric Effect

The photoelectric effect clarified the picture of light, exposing the particle-like side of its nature. The *photoelectric effect* refers to the observation that you can aim a beam of light at a sheet of metal, and that metal will emit electrons.

The photoelectric effect was explained in terms of light as waves, but two things puzzled physicists:

✔ Electrons were emitted immediately from the metal, even in low-intensity light (it was thought that light waves needed to build up the energy imparted to electrons).

✔ The kinetic energy of the emitted electrons was independent of the light intensity (it was thought that the more light, the more energy would be given to each emitted electron).

Albert Einstein, in a Nobel Prize–winning performance, explained both questions by introducing the idea of *photons* — that is, particles of light. Because each emitted electron was given its energy by a discrete light packet — a photon — electrons could be emitted from the metal as soon as light was shone on it. And because the electron's kinetic energy came from the photon, that kinetic energy was independent of the intensity of the light. For more on the photoelectric effect, flip to Chapter 13.

Davisson and Germer's Discovery of Matter Waves

The Davisson-Germer experiment confirmed the wave nature of electrons in 1927. That was quite a revolutionary discovery at the time, and it was a confirmation of the *de Broglie hypothesis* of matter waves. This hypothesis states that not only do waves sometimes behave like particles, but particles also sometimes behave like waves. For instance, a particle can be considered to have a wavelength related to its momentum.

In their experiment, Clinton Davisson and Lester Germer sent a beam of electrons onto a crystal of nickel. They proved that electrons that reflected from the highly smooth surface created an interference pattern — just as light waves would do — on a screen.

The electrons' wave-like nature was showing here: The crystalline structure of the nickel was diffracting the electrons, and the observed interference pattern was a sensational discovery — electrons behaved as waves!

Röntgen's X-rays

People can generate X-rays, the light rays that are so important in medicine, in vacuum tubes. You use a voltage to accelerate electrons to a very high speed; then the electrons hit a metal target and generate X-rays. Although such tubes were used in experiments earlier, German physics professor Wilhelm Conrad Röntgen was the first to report on X-rays in the late 19th century.

On November 8, 1895, Röntgen discovered X-rays when experimenting with such a vacuum tube. He was amazed by their penetrating power and their ability to produce clear images on photographic paper. He wrote a report titled "On a new kind of ray: A preliminary communication" in December 1895 and submitted it to the Würzburg Physical-Medical Society for publication. Röntgen got the very first Nobel Prize in Physics for the discovery.

Curie's Discovery of Radioactivity

In 1897, Marie Curie began her doctoral work and decided to investigate the "uranium rays" first discovered by Henri Becquerel. Using samples of radioactive substances, she and her husband, Pierre, eventually settled on *pitchblende* (a form of the mineral uraninite), which gave very strong photographic exposures through opaque paper. Eventually, they refined the pitchblende and discovered a new, radioactive element, *polonium,* named after Marie Curie's native Poland.

After chemically isolating the radioactive elements, the Curies observed that the elements were depleting while producing stable elements — principally helium and lead. They thereby discovered how the new "rays," superficially similar to the X-rays, were of a very different nature. These "rays" were a product of *radioactive decay* — the atomic process of the decay of unstable atoms into stable products.

The Curies also discovered a second radioactive element, *radium,* not long afterward. For her work, Marie Curie won not one but two Nobel Prizes, in Physics and Chemistry. You can read more about radioactivity in Chapter 15.

Rutherford's Discovery of the Atom's Nucleus

In the early 20th century, the reigning model of the atom was the English plum pudding model, which viewed the atom as a sort of positively charged paste in which electrons were embedded like plums.

Physicist Ernest Rutherford (who had already won a Nobel Prize in 1908 for his radioactivity work) dispelled that picture. In 1911, he aimed a beam of alpha particles at a thin gold foil. He discovered that contrary to plum pudding expectations, many alpha particles were scattered, even bouncing back entirely.

Rutherford said that the shock of observing the scattered alpha particles was like firing a "fifteen-inch shell" at tissue paper and having it come back and hit you. Clearly, positive charges were concentrated in the atom — electrons were too light to cause alpha scattering — and that was how the atomic nucleus was discovered. You can read more about this experiment in Chapter 14.

Putting a Spin on It: The Stern-Gerlach Experiment

In 1922, Otto Stern and Walther Gerlach conducted an experiment to determine whether particles had an intrinsic angular momentum. They set up a magnetic field in such a way that a stream of charged particles traveling through it would not be deflected unless they possessed at least a small *magnetic moment,* which quantifies the torque (turning force) experienced by a magnetic dipole moving through a magnetic field. The particles would possess a magnetic moment only if they had an intrinsic spin. And sure enough, the beam of particles split into two beams, indicating that charged particles (electrons in this case) did indeed have an intrinsic angular momentum, which Stern and Gerlach called *spin.*

This picture has significantly changed physicists' view of the electrons in atoms, because it adds another quantum number, *spin,* to each electron, doubling the number of electrons that can have the same other three quantum numbers. (The other quantum numbers — principal, orbital, and magnetic — specify orbital states of electrons in atoms.) Check out Chapter 12 for more info on quantum physics.

The Atomic Age: The First Atomic Pile

The first human-made, self-sustaining nuclear chain reaction occurred in 1942. Physicists started the reaction on December 2 of that year, beneath the west stands of Stagg Field, Chicago, beside a huge stack of carbon and uranium bricks comprising an *atomic pile* (the term people used before someone coined *nuclear reactor*). The time was 3:25 p.m. As *control rods,* which dampened the reaction, were withdrawn, the level of activity in the pile increased and held.

The atomic age was born. Since that time, interestingly, scientists found that Mother Nature beat us to the punch — not just with nuclear reactions inside stars, which are well-known, but also here on Earth. Self-sustaining nuclear reactions have been found in natural uranium deposits: Mother Nature's own atomic piles.

Verification of Special Relativity

Albert Einstein's theory of special relativity makes many claims that on the face of it seem pretty outlandish — length contraction? Time dilation? (See Chapter 12 for details.) But such effects have been borne out by experiment.

Take, for example, a mu meson, or *muon* for short. You find this particle in cosmic rays and in particle accelerators like those at CERN, a lab near Geneva, Switzerland. Muons have a very short lifetime — about one millionth of a second, so they're not around very long before they decay. On the other hand, subatomic particles can travel very fast — far faster than humans have been able to go so far — and you can observe relativistic effects when they do.

In particular, muons traveling at very high speeds last a lot longer than they should given their short lifetimes. That's because compared to the lab frame of reference, time really is dilated for the muons. Bruno Rossi and David Hall first observed the dilated lifetimes of muons in 1941, and many other experiments have also confirmed special relativity in great detail.

Chapter 17

Ten Online Problem-Solving Tools

In This Chapter
▶ Using online calculators
▶ Finding energy, reactance, frequency, half-life, and more

*P*hysics requires a lot of number crunching, and you can find help for that online. Many specialized physics calculators are available, and this chapter takes a look at some of the best. Just plug in your numbers, and the calculator can add your vectors, calculate frequency and wavelength, and even give you some quick numbers on relativity or radioactive decay.

Vector Addition Calculator

Vector addition can be very time-consuming. What's the direction of the net force from three charges on a test charge? What's the force's magnitude?

Now you can get some help with the vector addition calculator. Just enter the magnitude and direction (in degrees) of up to ten vectors, click the Calculate button, and you're there — the vector sum is displayed in two text boxes: One holds the vector sum's magnitude, and the other holds its direction. Simple.

You can find the vector addition calculator at

www.1728.com/vectors.htm

Centripetal Acceleration (Circular Motion) Calculator

If you have an electron orbiting in a magnetic field (see Chapter 4), you can calculate its centripetal acceleration using a lot of math — or you can let the centripetal acceleration calculator do it for you.

Select what you want to calculate — centripetal acceleration, radius, or velocity — from a drop-down box and then enter the other two values. The calculator does the rest.

You can find the centripetal acceleration calculator at

```
easycalculation.com/physics/classical-physics/
centripetal-acceleration.php
```

Energy Stored in a Capacitor Calculator

This calculator gives you the energy stored in a capacitor. You enter the capacitance in farads and the charge in coulombs; then click the Calculate button. The calculator displays the stored energy in the capacitor in joules.

You can find this calculator at

```
easycalculation.com/physics/electromagnetism/
stored-energy-electrical.php
```

Electrical Resonance Frequency Calculator

When you have a circuit with an inductor, a capacitor, and a voltage source that alternates at a given frequency, you can have *resonance* if you tune the frequency just right — that is, you can find a frequency that maximizes current in the circuit because the inductive reactance and the capacitive reactance cancel each other out (see Chapter 5 for details).

Now you can use an online calculator to find the resonance frequency of a circuit. You can also solve for the capacitance needed for resonance (given a voltage source frequency and an inductance), or you can solve for the needed inductance (given a voltage source frequency and a capacitance).

Just click the button indicating what you want to solve for — resonance frequency, capacitance, or inductance — click buttons to indicate which units you're using for each measurement (for instance, click either Henrys or MilliHenrys for inductance), enter the two numbers you know, and click the Calculate button; the number you want to solve for is displayed. Cool.

You can find the resonance frequency calculator at

```
www.1728.com/resfreq.htm
```

Capacitive Reactance Calculator

This calculator lets you figure the capacitive reactance of a capacitor, given a certain capacitance and a frequency. Just enter the two values and click the Calculate button, and you're done.

You can find this calculator at

    easycalculation.com/physics/electromagnetism/
    capacitive-reactance.php

Inductive Reactance Calculator

This calculator lets you calculate the inductive reactance of an inductor. You enter the inductance values and the frequency of the voltage source, click the Calculate button, and presto! The inductive reactance appears in a text box.

This calculator is at

    easycalculation.com/physics/electromagnetism/
    inductive-reactance.php

Frequency and Wavelength Calculator

This calculator lets you convert from frequency to wavelength or wavelength to frequency for light. Just enter a value in the Input box and click one of the buttons:

✔ **Buttons if you know wavelength:** cm, feet, meters

✔ **Buttons if you know frequency:** Hz, KHz, MHz

The calculator displays the corresponding value — for example, if you enter a value and click Hz, the calculator takes the value you entered as a frequency in hertz and displays the corresponding wavelength.

This calculator is at

    www.1728.com/freqwave.htm

Length Contraction Calculator

When you have speeds near the speed of light, you get length contraction. You can figure out what the length contraction is with the length contraction calculator. Just enter the fraction of the speed of light you're going (as a decimal) and click the second box. The length contraction factor appears in that box. Simple.

Find this calculator at

hyperphysics.phy-astr.gsu.edu/hbase/relativ/tdil.html

You can check out Chapter 12 for more info on length contraction and Einstein's theory of special relativity.

Relativity Calculator

The online relativity calculator specializes in calculations involving the relativity factor:

$$\frac{1}{\left(1 - \dfrac{v^2}{c^2}\right)^{1/2}}$$

The calculator changes units and solves for velocity or the relativity factor as you like. You enter a value in the Input box and then click one of these buttons:

- **Miles/second:** The calculator finds the relativity factor.
- **Kilometers/second:** The calculator finds the relativity factor.
- **c = 1:** The calculator finds the relativity factor using your input as a fraction of c.
- **Factor of change:** The calculator finds the speed needed to give you the input relativity factor.

You can find the relativity calculator at

www.1728.com/reltivty.htm

Half-Life Calculator

Working with radioactive decay is always a little tricky. Given a beginning amount of material and a half-life, how much material is left after a certain time? (*Half-life* is the time required for the amount of radioactive material to reduce by half through radioactive decay — see Chapter 15 for details.)

You can use the half-life calculator to find out. Click a button depending on what you want to solve for:

- Time (years)
- Half-life (years)
- Beginning amount (grams)
- Ending amount (grams)

Then enter the numbers as prompted. You can find the half-life calculator at

    www.1728.com/halflife.htm

Index

Business/Accounting & Bookkeeping

Bookkeeping For Dummies
978-0-7645-9848-7

eBay Business
All-in-One For Dummies,
2nd Edition
978-0-470-38536-4

Job Interviews
For Dummies,
3rd Edition
978-0-470-17748-8

Resumes For Dummies,
5th Edition
978-0-470-08037-5

Stock Investing
For Dummies,
3rd Edition
978-0-470-40114-9

Successful Time
Management
For Dummies
978-0-470-29034-7

Computer Hardware

BlackBerry For Dummies,
3rd Edition
978-0-470-45762-7

Computers For Seniors
For Dummies
978-0-470-24055-7

iPhone For Dummies,
2nd Edition
978-0-470-42342-4

Laptops For Dummies,
3rd Edition
978-0-470-27759-1

Macs For Dummies,
10th Edition
978-0-470-27817-8

Cooking & Entertaining

Cooking Basics
For Dummies,
3rd Edition
978-0-7645-7206-7

Wine For Dummies,
4th Edition
978-0-470-04579-4

Diet & Nutrition

Dieting For Dummies,
2nd Edition
978-0-7645-4149-0

Nutrition For Dummies,
4th Edition
978-0-471-79868-2

Weight Training
For Dummies,
3rd Edition
978-0-471-76845-6

Digital Photography

Digital Photography
For Dummies,
6th Edition
978-0-470-25074-7

Photoshop Elements 7
For Dummies
978-0-470-39700-8

Gardening

Gardening Basics
For Dummies
978-0-470-03749-2

Organic Gardening
For Dummies,
2nd Edition
978-0-470-43067-5

Green/Sustainable

Green Building
& Remodeling
For Dummies
978-0-470-17559-0

Green Cleaning
For Dummies
978-0-470-39106-8

Green IT For Dummies
978-0-470-38688-0

Health

Diabetes For Dummies,
3rd Edition
978-0-470-27086-8

Food Allergies
For Dummies
978-0-470-09584-3

Living Gluten-Free
For Dummies
978-0-471-77383-2

Hobbies/General

Chess For Dummies,
2nd Edition
978-0-7645-8404-6

Drawing For Dummies
978-0-7645-5476-6

Knitting For Dummies,
2nd Edition
978-0-470-28747-7

Organizing For Dummies
978-0-7645-5300-4

SuDoku For Dummies
978-0-470-01892-7

Home Improvement

Energy Efficient Homes
For Dummies
978-0-470-37602-7

Home Theater
For Dummies,
3rd Edition
978-0-470-41189-6

Living the Country Lifestyle
All-in-One For Dummies
978-0-470-43061-3

Solar Power Your Home
For Dummies
978-0-470-17569-9

Internet

Blogging For Dummies,
2nd Edition
978-0-470-23017-6

eBay For Dummies,
6th Edition
978-0-470-49741-8

Facebook For Dummies
978-0-470-26273-3

Google Blogger
For Dummies
978-0-470-40742-4

Web Marketing
For Dummies,
2nd Edition
978-0-470-37181-7

WordPress For Dummies,
2nd Edition
978-0-470-40296-2

Language & Foreign Language

French For Dummies
978-0-7645-5193-2

Italian Phrases
For Dummies
978-0-7645-7203-6

Spanish For Dummies
978-0-7645-5194-9

Spanish For Dummies,
Audio Set
978-0-470-09585-0

Macintosh

Mac OS X Snow Leopard
For Dummies
978-0-470-43543-4

Math & Science

Algebra I For Dummies,
2nd Edition
978-0-470-55964-2

Biology For Dummies
978-0-7645-5326-4

Calculus For Dummies
978-0-7645-2498-1

Chemistry For Dummies
978-0-7645-5430-8

Microsoft Office

Excel 2007 For Dummies
978-0-470-03737-9

Office 2007 All-in-One
Desk Reference
For Dummies
978-0-471-78279-7

Music

Guitar For Dummies,
2nd Edition
978-0-7645-9904-0

iPod & iTunes
For Dummies,
6th Edition
978-0-470-39062-7

Piano Exercises
For Dummies
978-0-470-38765-8

Parenting & Education

Parenting For Dummies,
2nd Edition
978-0-7645-5418-6

Type 1 Diabetes
For Dummies
978-0-470-17811-9

Pets

Cats For Dummies,
2nd Edition
978-0-7645-5275-5

Dog Training For Dummies,
2nd Edition
978-0-7645-8418-3

Puppies For Dummies,
2nd Edition
978-0-470-03717-1

Religion & Inspiration

The Bible For Dummies
978-0-7645-5296-0

Catholicism For Dummies
978-0-7645-5391-2

Women in the Bible
For Dummies
978-0-7645-8475-6

Self-Help & Relationship

Anger Management
For Dummies
978-0-470-03715-7

Overcoming Anxiety
For Dummies
978-0-7645-5447-6

Sports

Baseball For Dummies,
3rd Edition
978-0-7645-7537-2

Basketball For Dummies,
2nd Edition
978-0-7645-5248-9

Golf For Dummies,
3rd Edition
978-0-471-76871-5

Web Development

Web Design All-in-One
For Dummies
978-0-470-41796-6

Windows Vista

Windows Vista
For Dummies
978-0-471-75421-3

Available wherever books are sold. For more information or to order direct: U.S. customers visit www.dummies.com or call 1-877-762-2974.
U.K. customers visit www.wileyeurope.com or call (0) 1243 843291. Canadian customers visit www.wiley.ca or call 1-800-567-4797.

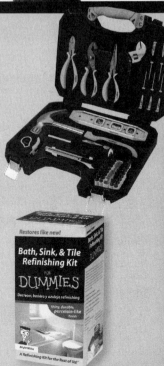